Key Concepts in Geomorphology

Key Concepts in Geomorphology

Daniel Hunter

STATES
ACADEMIC PRESS
www.statesacademicpress.com

States Academic Press,
109 South 5th Street,
Brooklyn, NY 11249, USA

Visit us on the World Wide Web at:
www.statesacademicpress.com

ISBN: 978-1-63989-315-7 (Hardback)

Cataloging-in-Publication Data

Key concepts in geomorphology / Daniel Hunter.
p. cm.
Includes bibliographical references and index.
ISBN 978-1-63989-315-7
1. Geomorphology. 2. Physical geography. 3. Landforms. I. Hunter, Daniel.
GB401.5 .K49 2022
551.41--dc23

Table of Contents

	Preface	**VII**
Chapter 1	**Introduction to Geomorphology**	**1**
	• Branches of Geomorphology	13
Chapter 2	**Concepts and Models in Geomorphology**	**70**
	• General Systems Theory	70
	• Hardware Models in Geomorphology	84
	• Mathematical Model	97
	• Experimental Design Models	108
Chapter 3	**Understanding Primary Geomorphic Processes**	**110**
	• Aeolian Features and Processes	110
	• Fluvial Processes	125
	• Glacial Processes	135
	• Hillslope Processes	153
Chapter 4	**Types of Weathering**	**163**
	• Physical Weathering	165
	• Chemical Weathering	167
	• Biological Weathering	171
Chapter 5	**Landforms in Geomorphology**	**172**
	• Aeolian Landforms	175
	• Glacial Landforms	185
	• Fluvial and Coastal Landforms	197
	• Volcanic and Slope Landforms	221
	Permissions	
	Index	

Preface

The branch of science which is concerned with the origin and evolution of various landscapes on Earth is known as geomorphology. This includes the study of various chemical, physical and biological processes on Earth like air, ice and water, and their effects on the formation of topographic and bathymetric features. Geomorphology maps the distribution of different types of landforms such as sand dunes, glaciers, etc. Some of the important processes that are studied under this discipline are aeolian processes, biological processes, fluvial processes, glacial processes, igneous processes, hillslope processes and tectonic processes. The subject of geomorphology often overlaps with various other fields like glaciology, civil engineering, environmental engineering, sedimentology, etc. It finds extensive application in a wide variety of areas such as river control, stream restoration, hazard assessment and coastal protection. This book is a valuable compilation of topics, ranging from the basic to the most complex theories and principles in the field of geomorphology. It studies, analyzes and upholds the pillars of geomorphology and its utmost significance in modern times. The book is appropriate for those seeking detailed information in this area.

To facilitate a deeper understanding of the contents of this book a short introduction of every chapter is written below:

Chapter 1- The scientific study of the origin and classification of the Earth's topographic and bathymetric features is known as geomorphology. Some of the branches of geomorphology are biogeomorphology, urban geomorphology, climatic geomorphology, hydrogeomorphology, soil geomorphology and applied geomorphology. This is an introductory chapter which will briefly introduce all these significant aspects of geomorphology.

Chapter 2- Models in geomorphology are used for the conceptual and quantitative representations of the surface of the Earth. There are various models which are used in geomorphology such as hardware models, mathematical model and experimental design models. This chapter has been carefully written to provide an easy understanding of these models in geomorphology.

Chapter 3- Geomorphic processes refer to the endogenic and exogenic forces that cause physical stress and chemical actions to bring about changes in the configuration of the Earth's surface. Some of the different types of geomorphic processes are fluvial processes, glacial processes, hillslope processes, igneous processes and marine processes. This chapter discusses in detail these various types of geomorphic processes.

Chapter 4- The process of breaking down of rocks and minerals on the surface of the Earth by water, atmospheric gases and biological organisms is termed as weathering. It can be categorized into physical, chemical and biological weathering. The topics elaborated in this chapter will help in gaining a better perspective about weathering.

Chapter 5- The features created by the erosive and constructive actions of the wind are known as aeolian landforms whereas the landforms produced by the action of glaciers are termed as glacial landforms. This chapter closely examines the key concepts of aeolian, glacial and fluvial processes and landforms to provide an extensive understanding of the subject.

I would like to share the credit of this book with my editorial team who worked tirelessly on this book. I owe the completion of this book to the never-ending support of my family, who supported me throughout the project.

Daniel Hunter

Chapter 1

Introduction to Geomorphology

The scientific study of the origin and classification of the Earth's topographic and bathymetric features is known as geomorphology. Some of the branches of geomorphology are biogeomorphology, urban geomorphology, climatic geomorphology, hydrogeomorphology, soil geomorphology and applied geomorphology. This is an introductory chapter which will briefly introduce all these significant aspects of geomorphology.

Geomorphology is the study of landforms, their processes, form and sediments at the surface of the Earth (and sometimes on other planets). Study includes looking at landscapes to work out how the earth surface processes, such as air, water and ice, can mould the landscape. Landforms are produced by erosion or deposition, as rock and sediment is worn away by these earth-surface processes and transported and deposited to different localities. The different climatic environments produce different suites of landforms. The landforms of deserts, such as sand dunes and ergs, are a world apart from the glacial and periglacial features found in polar and subpolar regions. Geomorphologists map the distribution of these landforms so as to understand better their occurrence.

Earth-surface processes are forming landforms today, changing the landscape, albeit often very slowly. Most geomorphic processes operate at a slow rate, but sometimes a large event, such as a landslide or flood, occurs causing rapid change to the environment, and sometimes threatening humans. So geological hazards, such as volcanic eruptions, earthquakes, tsunamis and landslides fall within the interests of geomorphologists. Advancements in remote sensing from satellites and GIS mapping has benefited geomorphologists greatly over the past few decades, allowing them to understand global distributions.

Geomorphologists are also "landscape-detectives" working out the history of a landscape. Most environments, such as Britain and Ireland, have in the past been glaciated on numerous occasions, tens and hundreds of thousands of years ago. These glaciations have left their mark on the landscape, such as the steep-sided valleys in the Lake District and the drumlin fields of central Ireland. Geomorphologists can piece together the history of such places by studying the remaining landforms and the sediments – often the particles and the organic material, such as pollen, beetles, diatoms and macrofossils preserved in lake sediments and peat, can provide evidence on past climate change and processes.

So geomorphology is a diverse discipline. Although the basic geomorphological principles can be applied to all environments, geomorphologists tend to specialise in one or two areas, such aeolian (desert) geomorphology, glacial and periglacial geomorphology, volcanic and tectonic geomorphology, and even planetary geomorphology. Most research is multi-disciplinary, combining the knowledge and perspectives from two contrasting disciplines, combining with subjects as diverse as ecology, geology, civil engineering, hydrology and soil science.

Contemporary geographers classify geomorphology (as a branch of geography) under the Earth Science Group. The present figure demonstrates that several other branches of knowledge are directly and indirectly linked with geomorphology. These allied disciplines are having a kind of mutual relationship with geomorphology. For example, according to Spark the main contribution of geomorphology to geology is in unraveling of geographical successions, for example when deposits are thin and scattered such conditions being most characteristic of Pliocene and Pleistocene periods in certain areas.

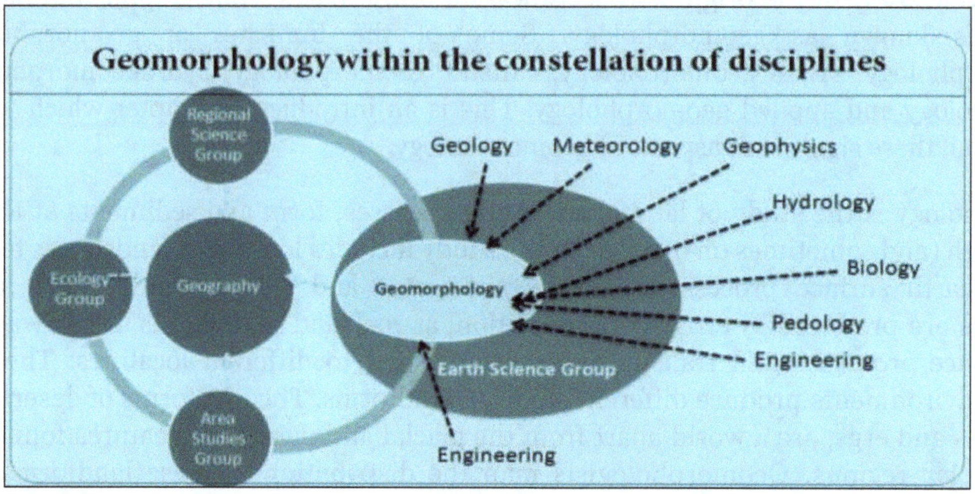

Similarly geomorphologists must have good working knowledge of geology since without knowing lithology of rock geomorphologist cannot fully understand the formation and character of landform. One of the fundamental principles of geomorphology also states that "Geological structure is a dominant control factor in the evolution of land forms and is reflected in them". Likewise meteorology provides information related with effect of climatic elements on the rate and nature of geomorphic processes. In turn geomorphology contributes to meteorology by disclosing effect of morphology on climatic variables. For example, the rainfall during the Monsoon season is largely affected by the orographic effect of Himalayas. On the other hand it is equally true that "an appreciation of world climates is necessary to a proper understanding of the varying importance of the different geomorphic processes."

Recent Trends in Geomorphology

The present day geomorphologists are largely influenced by two important revolutions occurred in the discipline of Geography i.e. 'Quantitative revolution' and the 'Process revolution'. The former introduced explicit emphasis on the use of scientific methods and later developed the greater understanding of the processes responsible for the creation of variable character of the earth surface.

The measurement, monitoring, analysis and modelling of the formative processes have become integral part of present day geomorphology. For example, Differential GPS (or DGPS) is now commonly used by geomorphologists to identify precise location of topographical features. This prices measurement is also extremely useful for the changes happening in the glacier areas especially to monitor climate change. It is equally helpful for the identification of hazards related with landslides and slope failures. The Ground based Light Detection and Ranging (LIDAR) is another instrument to identify precise location of third order landforms especially in inaccessible areas of the earth. The modern day sensitive instruments can even track the relative movement of different plates, glaciers as well as lava flow in a volcanic region.

The high resolution satellite imageries, Computer aided Cartography, Electronic Distance Measurement (EDM), Global Positioning System (GPS) and GIS softwares have immensely contributed to the development of a new specialized branch of geomorphology i.e., Geomorphometry.

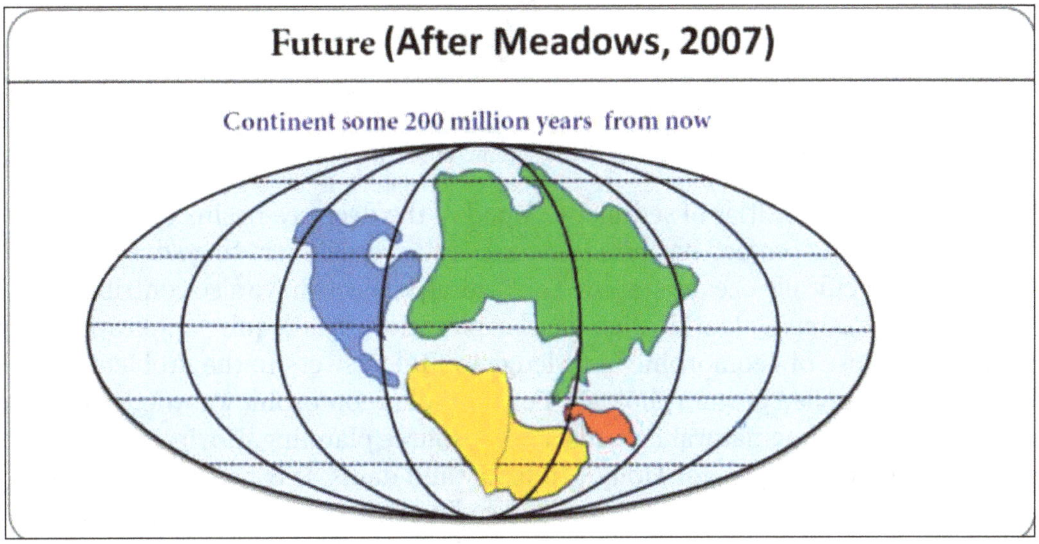

Now within a fraction of a second gemorphologists can compute topographic elements through Digital Elevation Models (DEMs) and Digital Terrain Models (DTMs).

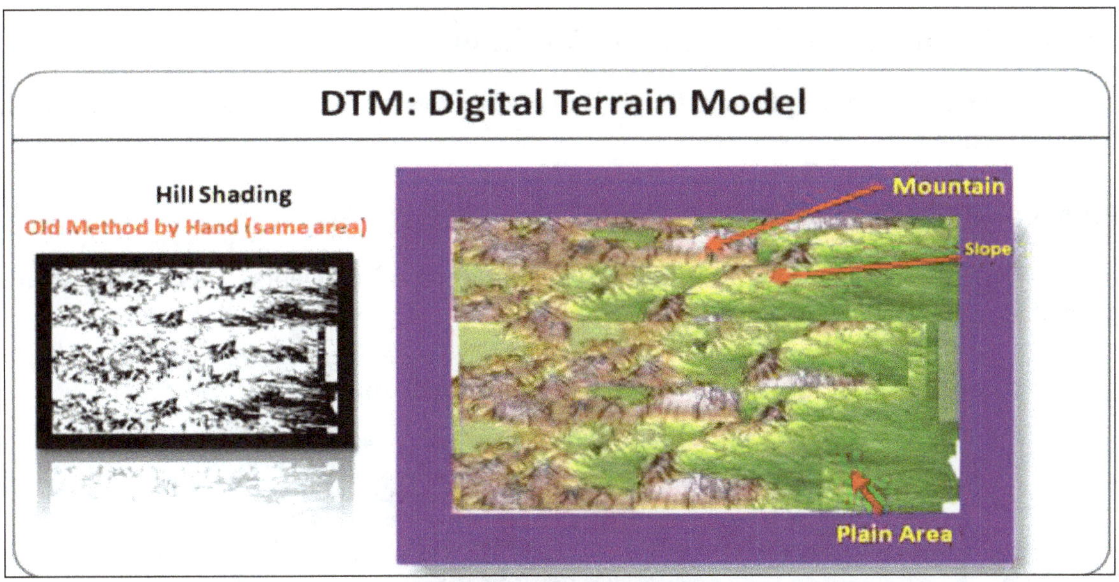

The freely available Google Earth satellite imageries have given new tool to explore the inaccessible parts of the world. Now a volcanologist can virtually explore an active volcano through the near real time satellite imageries. For example the present imagery is showing an aerial view of a crater at the summit of Mt St. Helens (USA). This image has been taken by satellite.

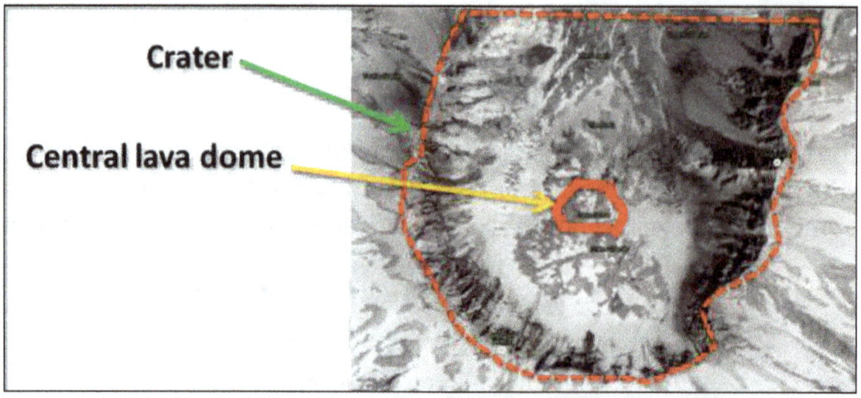

Aerial view of a crater at the summit of Mt St. Helens (USA).

A wealth of data on the quantities of sediments found in the deep sea basins of the planet earth by 'Integrated Ocean drilling project' has lured several young researchers to go deep into the field of geomorphology or specifically oceanography. The drilling projects have also contributed to applied geomorphology. For example identification of oil exploration sites require application of geomorphic knowledge. The use of geomorphic knowledge to find answers to the problems faced by human beings is called 'applied geomorphology'. For example geomorphic knowledge is prerequisite for managing and preventing natural hazards, land resource planning, Environmental Impact Assessment, site suitability analysis for House, Industry and dams. It is more advantageous that we know, for instance, something of the discharge and sediment load of streams, than they are 'young' or 'mature', as Davis described them. We will also attempt to discuss applied geomorphology in separate module of this paper. At present with the advent of the internet and improved access to many sources, the research literature in geomorphology has increased rapidly.

The State of Geomorphology: SWOC Analysis

Let us see the state of geomorphology through SWOC analysis i.e. Strength, Weaknesses, opportunities and challenges:

Strength and Opportunities

Geomorphology is enriched with well-developed fundamental concepts. There is no dearth of books related to geomorphology.

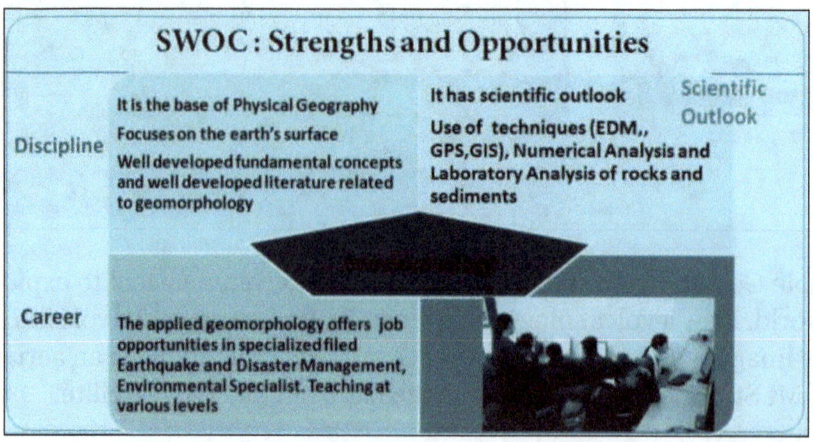

Geomorphology is scientific in nature. The techniques like EDM, GPS and GIS, Numerical Analysis and Laboratory analysis of rocks and sediments, and several instruments used in filed investigations makes scientific identity of the discipline.

Several developments in the field of science and technology has opened many specialized branches of geomorphology such as, remote-sensing geomorphology, experimental geomorphology, extra-terrestrial (planetary) geomorphology, engineering geomorphology and anthropogenic-geomorphology. One can also become an expert in these branches. For example, Fluvial Geomorphologist, Karst Geomorphologists, Coastal Geomorphologists, Biogeomorphologist, Tropical Geomorphologist, Mountain Geomorphologist. The applied geomorphology also offers many job opportunities in specialized fields. For example expert in disaster management EIA expert and Environmental Specialist.

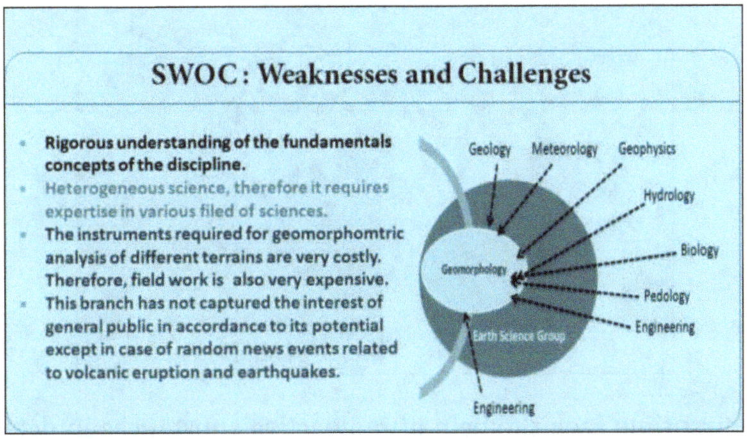

Weaknesses and Challenges

Geomorphology requires a rigorous understanding of the fundamentals concepts of the discipline. It is basically a heterogeneous science; therefore it requires expertise in various filed of sciences. The instruments required for geomorphometric analysis are very costly. Therefore, the field work is very expensive. This branch of physical geography has also not captured the interest of general public except in case of random news events related to volcanic eruptions and earthquakes. Recent massive 7.9 magnitude shallow focus Nepal earthquake on 25, April, 2015 at 11.48 AM (IST) has shocked the world. It claimed the life of more than 9,000 people in our neighbouring country Nepal, which is one of the worst in Nepal's history. Nepal is still coping with the aftermath of the earthquake. Therefore geomorphic literacy is extremely important not only to understand the phenomena but also for disaster.

Geological Time Scale

Life on the earth did not come into being with its birth. The earth has a long history and has seen many changes on her surface through different periods. The earth cooled from its gaseous state into liquid one and transformed into solid layer on further cooling. For an instance, there was a great ocean in place of the Himalayas only 150 million years back. With one cellular organism, life began which belonged to both animal and vegetation kingdom. Later on, the animals and vegetations both evolved into more complex life-forms. The animals evolved into fish, reptiles, mammals and finally into human beings. The great age of the earth were based on the rate of sedimentation, age of fossils, and the rate of cooling of the earth, salinity of the oceans and the tidal force of the moon. As far as, the age of the Earth is concerned, there is a lot of difference of opinion among the scientists. On the basis of current methods of radio-metric dating, most of the modern scientists agree that the life of the Earth may range from 3 to 5 billion years.

James Hutton believed in the principle of uniformitarianism and he knew that the earth was very old but they had no method of knowing its exact age. Several methods were devised by others which were in use till about the beginning of the 20th century. Lord Kelvin, a renowned physicist, had assumption that the earth was originally molten and cooled to its present condition, put the earth's age at not more than 100 million years, and perhaps much less.

Rutherford, for the first time took the help of radioactive substances in determining the age of the earth. For calculating the ages of rocks and minerals, the discovery of radio-activity provided a reliable means; contain radioactive isotopes (elements). This method is called radiometric dating. Till about 40 years ago the age of the earth calculated on this basis was placed at about 3,000 million years. In recent years, it has been possible to make more accurate calculations due to more precise scientific knowledge. The relative percentage of lead isotopes in ancient rocks and meteorites has also been taken into account and the age of the earth is now placed at around 4,600 million years. Chances of error in this calculation are very small; therefore this age is now widely accepted.

Geological History of the Earth

Though life on earth has existed only for a small fraction of her history, fossils, which are remains of extinct animals, indicate a good deal. It is possible to estimate the depth of the sea on the basis of fossils of certain marine animals like corals, where particular sediment was precipitated. Furthermore fossils provided indications of past environments. Radiometric dating has made possible the specific or absolute dating of rock units that represent various events in the earth's distant past. When a complete sequence of rocks representing the entire period is not available and there are breaks in the rock record, then it is known unconformities.

The principal divisions of geological time are called eras. Three eras are recognised- the Plaeozoic (ancient life), the Mesozoic (middle life) and the Cenozoic (recent life). Each of these eras is subdivided into smaller time units known as periods.

The Palaeozoic has seven periods, Mesozoic and Cenozoic has three and two periods respectively. Each period is characterized by its profound changes in life forms. These twelve periods is further sub-divided into epochs and epoch into ages. It will be observed that the Palaeozoic era begins some 600 million years ago. The enormously long period of more than 4 billion years prior to the Pre-cambrian with no further subdivision, as our knowledge about those early times is very limited. This is so, because prior to the Cambrian, very simple life forms such as algae, bacteria, fungi, sponges and worms have not left adequate fossil record. On the basis of the above discussion a standard geological timescale has been prepared which is accepted throughout the world and with reference to which the geological history of the earth can be studied.

Stratigraphic Chart

Era	Period	Epoch	Million YearsAge	Distinctive Life	Major Structural Episodes	
Cenozoic	Quaternary	Recent/Holocene	0.01	Ice age ends Humans are dominant		
		Pleistocene/ more recent	2.5	Age of Man	Main Episode	Alpine
	Tertiary	Pliocene	12	Age of Mammals- (flowering plants and broad leaved trees)		
		Miocene	25			
		Oligocene	35-40		Laramide Phase	
		Eocene	60			
		Paleocene	70			
Mesozoic	Cretaceous (Creta=chalk)		135	Age of reptiles (Dinosaurs) and birds	General Worldwide Elevation	
	Jurassic(after Jura mountain)		180			
	Triassic (three fold division in Germany)		225			
Palaeozoic	Permian		270	Age of Amphibians	Main Episode	Hercynian
	Carboniferous		350	Widespread forests		
	Devonian		440+101	Age of fishes		
	Silurian		440+10	Age of marine		
	Ordovician		500	Invertebrates		
	Cambrian		600	Abundant fossils first appear	Main Caledonian Episode	
Azoic or Precam-brian			Oldest rocks 3787+85		Several Mountain Building Episodes	

It is clear from the above table that man is comparatively a new comer on the Earth. Even this history of the earth constitutes not more than one sixth of the life of the earth, nothing is known about 5/6th of the Earth's history.

Division of Geological Time into Eras

We have noticed above that geological time has been divided into four major divisions called eras-Pre-cambrian, Palaeozoic, Mesozoic and Cainozoic or Cenozoic. The oldest and by far the longest is Pre-Cambrian which covers 90 per cent of geological time but our knowledge about this era is too little.

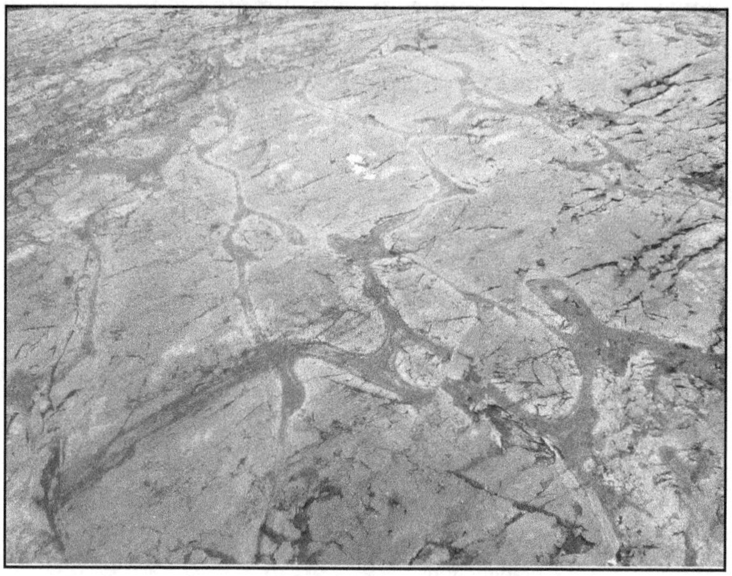

Weathered Precambrian pillow lava in the Temagami Greenstone Belt of the Canadian Shield.

Pre-Cambrian Era (Eozoic)

This is the oldest era of geological history. The duration of this era is from the beginning of the earth nearly 4.6 billion years or more till about 0.6 billion years ago. In other words Pre-Cambrian alone compasses 90 per cent of all geological time.

Pre-Cambrian rocks are, which are said to be the oldest one, belong to the Archaean period. Most part of the Peninsular plateau of India is composed of Archaean rocks which include the rocks of the Dharwar system which were originally sedimentaries. Just above the Archaean system are the Late Pre-Cambrian or Proterozoic rocks, the rocks of the Vindhyansystem and the Cuddapah system belong to this period. The rocks of the Pre-Cambrian era cover extensive areas in all the continents and usually form areas of low or moderate relief on account of the denudation of ages: These are called continental shields. In this era, Pangaea continent was born and at the end of the period the four shields had come into existence- i.e. Laurentian shield; Siberian shield; Gondwana shield; Baltic shield.

The Pre-Cambrian rocks are extremely rich in metallic minerals, and almost all the important occurrences of iron ore, gold, copper, manganese, uranium, chromium, lead, zinc and mica in the world are found in the rock of this era.

Paleozoic Era

The Palaezoic started 600 million years ago and continued till about 225million years. Its time range is about 30 million years. The animals in the seas and oceans were invertebrate. Non-floral vegetation came into existence. The chief rocks of this period are shale, slate, hard sandstone, hard limestone, and rocks of this period are found large deposit of valuable minerals. The rocks of this period in North America abound in gold, oil, coal and iron ore while in Europe tin, copper and iron are found in the rocks of this period. This era further sub-divided into six periods of which the oldest one is the Cambrian followed by the Ordovician, Silurian, Devonian, Carboniferous and the Permian.

During this era, the South Pole was possibly somewhere in the South Atlantic Ocean and the North Pole in the North Pacific Ocean and the equator passed through South Australia, Arabian Peninsula, Middle Europe and eastern North America. In the Devonian period presence of coral a lime stone in North America, Europe and South-east Asia and Australia indicates a hot climate. On the other hand, in the Devonian and Permian periods there are evidences of relatively cold climates in South America, South Africa and India.

1. Cambrian Period: This period was about 90 million yearlong (from 600 to 500 million years). Cambrian Period saw the transgression of sea over the land surface. Land submerged under water due to earth movement that gave rise to sedimentary rocks. It is why the rocks of this period are quite extensive. The rocks of this time have light limestone at the bottom over which are laid the layers of sandstone, shale and at the top is a thin layer of limestone again. These rocks not only contain the fossils of the oldest life but also hold the first evidence of life.

2. Ordovician Period: The period witnessed a large portion of the land surface being submerged under water and half of North America was under water. There was an expansion of oceans in this period. Several volcanic eruptions took place in this period and the eastern portion of the U.S.A. was overlaid with the volcanic ash and dust. Among the animals the greatest evolution of Trilobites took place. The sea-grass was the main vegetation with characteristic of the birth of molluscus in this period. Slates, lime stones and volcanic rocks spread over a large area.

3. Silurian Period: It was the third period and followed the Ordovician Period and preceded the Devonian Period. This period was important for the birth of fish and land vegetation. The Caledonian earth movement affected all the continents. For the first time, the animals which could breathe came into existence that means animals with lungs developed. Fish came into existence. The red sandstone also came to be formed in this very period.

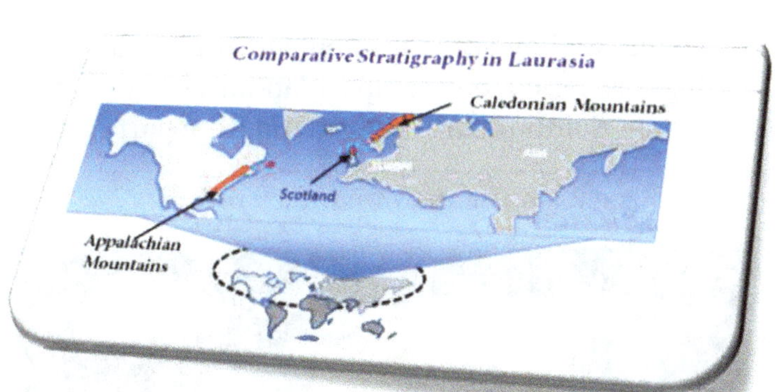

4. Devonian Period: It was about a million year long period. It began 440 million years ago and ended 350 million years ago. Devonian Period saw at its beginning the well-known mountain building movement i.e. Caledonian movement. The mountains of Scandinavia, North England,

Scotland and Spitbergen were formed in this period. It is called the age of fishes because the seas were full of them. In the late Devonian, the first amphibians and vertebrate colonisation of the dry land had begun. All the rocks of this system hold the fossils of fish. The chief creatures of this period were predominantly aquatic.

5. Carboniferous Period: It was 80 million year long. It began 350 million years ago and ended in 270 million years ago. The importance of this period is that the coal beds of the earth came to be formed, hence it is also known as the Coal Age. Thus in course of time due to intense heat and pressure, the successive layers of decaying vegetation, entrapped in the sediment layers, were transformed into coal beds. This period also saw the beginning of such animal life which could live both on land and water.

6. Permian Period: This period is 45 million years long. It began 270 million years ago and ended 225 million years ago. Permian Period was the time when the Hercynian mountain building movement occurred. This movement left its mark in central Europe, North America and several other places. There was a heavy deposition of red rocks and lime stones in South-West U.S.A. The animals which could live both on land and sea came to evolve all the more and reptiles saw the light of day. The temperature began to rise and the climate started getting dry in this period.

Mesozoic Era

Mesozoic or secondary Era is formed of two words-meso i.e. middle and zoic i.e. life. In other words, it is the meeting time between the old and the new Eras. The Mesozoic era is of a much shorter duration than the Palaeozoic. The climate started getting warm and dry. The snow cover of Antarctica began to melt. This era is best known as the age of dinosaurs and reptiles. In this era, birds make their appearance in Jurassic and the small mammals begin to appear in the Triassic. The Tethys Sea was divided into two parts. This era consists of the following three periods:

- Triassic Period,
- Jurassic Period,
- Cretaceous Period.

1. Triassic Period: It started 225 million years ago and ended 180 million years ago. It was about 40 million year long. During this period, one compact landmass known as Gondwanaland; South and Central Africa, Madagascar, South India and Australia were part of it.

Laurasia and Gondwanaland

During this period, while the northern hemisphere was having dry climate, the southern hemisphere was under laid with an ice sheet. The Tethys Sea which had come into existence by the end of the Palaeozoic era, continued to exit till the Cretaceous without any significant deformation. The places in Europe where we now find the Mediterranean Sea and the Alps mountains, there used to flow the Tethys Sea.

The shales and sandstones of the Panchet series of the Gondwana system belong to the Triassic and the sandstones and shales of the Jabalpur series belong to the Jurassic. In this period; shale, limestone, sandstone, etc. were formed. The first dinosaur was born mammals also came into existence in the elementary form.

2. Jurassic Period: It is considered to be 45 million year long. It started 180 million years ago and ended 135 million years ago. Reptiles reach their greatest proportions in this period. They dominated the land, sea and air. The fossils of the first bird also belong to this period. A lot of deposition took place in the seas of western USA. Nevoda and Laramide movements also took place in Jurassic period. Formation of Sierra Nevada Mountains was the result of these movements.

3. Cretaceous Period: It began 135 million years ago and ended 70 million years ago. It was about 60 million year long. It is well-known for its widespread deposits of chalk. The other major rocks of this period are soft sandstone, clay and limestone. Due to Laramide movement, the American Cordilleras were formed. It was during this period that deciduous trees appeared. Undoubtedly, the most dominating among the animals were the reptiles. The first birds and mammals were also appeared in this period. Huge animals(Dinosaurs) began to decay and were almost extinct by the end of the period.

Cenozoic Era

This latest era in the earth's geological history has began about 65 million years ago. Cenozoic is made of two words-ceno, i.e. new and zoic, i.e. life. It is an era of new life which normally divided into two periods-Tertiary and Quaternary. But there are some people who are in favour to regard the Tertiary and Quaternary as two separate eras. Tertiary period ended about one million years ago. At present we are living in the Quaternary period.

The Tertiary period is sub-divided into five epochs. These are, beginning from the oldest, Paleocene, Eocene, Oligocene, Miocene and Pliocene. It was during this era that the vegetation increased and the ape man appeared.

- Paleocene Epoch: This epoch began 70 million years ago and ended 60 million years ago. It is considered to be about 10 million years long. The Rocky Mountains were formed under Laramide movement. Grass land area also increased in this epoch.
- Eocene Epoch: It is considered to be 20 million years long. It started 60 million years ago and ended 40 million years ago. The land area developed many fissures which belched out lava over a big area. Mammals evolved into many branches. During this epoch plants bearing fruits and grains developed on a large scale.
- Oligocene Epoch: It began 40 million years ago and ended 25 million years ago. It was about 15 million years long. It was the epoch of Alpine Orogeny. Sediment started depositing in the seas of Pacific coast. The forefather of man, Anthropoid, was born in this epoch.

- Miocene Epoch: It was almost 13 million years long. It began 25 million years ago and ended 12 million years ago. The rivers continued to deposit sediments in the Great-Plains of U.SA. During this epoch the Alpine folds developed in Eurasia and trended in the east-west direction. Sediment was also deposited at large scale in the Atlantic ocean and Gulf coast. The whale and monkey came into existence.
- Pliocene Epoch: This epoch was about 10 million years long. It started 12 million years ago and ended 2.5 million years ago. Due to orogenetic movement, the sediment deposited in the shallow seas rose in the form of plains. During this epoch large plains developed, the important ones are the Northern Plains, plains of Europe, the Po basin of Italy, the Mesopotamian plain, Ganga and Sindh plains. Horses, elephants and huge carnivorous animals evolved. This epoch is also considered to be the evolution of man.

The Quaternary period started about two and a half million years ago. It is further subdivided into two epochs- Pleistocene and the Recent (or Holocene). This is the most recent of the periods of geological history. The alluvial deposits of the world were also laid in this period. The rocks of this period contain the fossils of horses, elephants, camels and several other oceanic mammals. This period saw the evolution of man and the growth of his intellect.

1. Pleistocene Epoch: The Pleistocene Epoch is typically defined as the time period that began about one million years ago and lasted until about 10,000 years ago. Man appeared on this planet only towards the beginning of Pleistocene but the modern man developed in the Holocene epoch. At the beginning of this epoch, the temperature of the atmosphere became so low that the surface of the earth came under a vast ice sheet. Besides these, the Himalayas and the Alps were having small patches of ice-caps. On account of these, this era is also known as the Great lce Age. The land masses of the Southern Hemisphere, however, remained free from ice. North Africa, North America (Canada and northern U.S A.), Arctic and Antarctica were all covered with a thick layer of snow. The snow caps of high mountains expanded. The snow cover of North America has disappeared hardly 10,000 years ago. The snow cover had a prevalent effect upon human life. Large and violent animals were killed. Mammals had evolved very fast. Monkeys, chimpanzees and human beings increased in number at large. Man's fossils are found in rocks about one hundred thousand years old. It is, therefore, estimated that man is only one hundred thousand years old.

The snow cover was not static in this age. From a study, snow cover advanced and retreated four times. Hence this period is called ice age and divided into four ice-ages (a) Gunz (b) Mindel (c) Riss and (d) Wurm. There is no unanimity among the scientist regarding causes of ice-ages. Some of the causes have been attributed to the birth of mountains, the eruption of volcanoes, the interruptions in the receipt of solar energy, etc. The five Great Lakes of North America were formed by the retreat of the snow cover. The mountains tops of Europe and America were rounded and the river valleys developed U-shape. The landscapes features of Sweden, Finland and Russia are a testimony to the work of snow cover.

2. Holocene Epoch: This epoch started 10,000 years ago. 'Man' is the most important feature of the Holocene epoch that is why Holocene can be called the age of human-evolution. However, the theory of human evolution is still in question. Man, by the development of his intellect has established his control over the flora (plants) and fauna (animals) of the world. He has accelerated some changes in the landforms and has converted the Earth into a place for his existence. Though no

large scale changes have been effected by rivers, sea waves, winds, etc., in the landforms during the Holocene, yet man has made tremendous progress in human civilization.

During his relatively short period man by his intelligence, mental ability and power of organization has greatly speeded up the pace of development. Therefore the climatic conditions in the past have played an important role in the way of human life. The effects on the life of man are visible on the type of environment creating for his comfort. Man has also been able to move out of the pull of the Earth and visited the Moon and constant efforts may make human landing on Mars.

Branches of Geomorphology

Biogeomorphology

Biogeomorphology (or ecogeomorphology) of the two way interplay between organisms and geomorphology. Geomorphologists are primarily interested in the evolution and erosion of landforms across the globe including iconic geodiversity features such as coral reefs and sea stacks. Biogeomorphologists study how organisms shape the landscape and landforms and how their activity is mediated by other geomorphological processes.

Biogeomorphological Organisms can:

- Construct new landforms and are called bioconstructors. Corals, mangrove forests and beaver dams are good examples.
- Erode landforms and are called bioeroders. Sea urchins and badgers are good examples.
- Protect landforms from other agents of deterioration and erosion, such as waves and salts. These are called bioprotectors and ivy and barnacles are good examples.

Some species are mainly protective but sometimes cause deterioration (e.g. ivy) and the relationships between organisms and geomorphology are often complex.

The Work of Biogeomorphological Species

They are also called ecosystem engineering as plants, animals and microorganisms are directly or indirectly involved in sculpting the physical habitat of other organisms. For example, tiny, microscopic organisms (microns) have been found to drill into rock surfaces making them retain water for longer and increasing surface roughness. This then provides a wetter, rougher surface which makes it easier for animals such as barnacles to colonise and survive in the intertidal zone.

Biogeomorphology research is often applied to help solve real-world problems including saltmarsh and mangrove restoration; coastal structure designs; river restoration and building management and heritage conservation. Researchers often advise conservation and environment agencies on management issues where it is helpful to 'work with natural processes'.

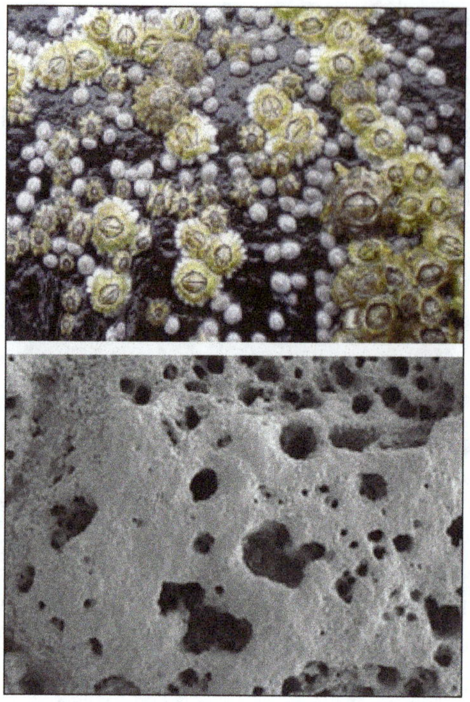

Biogeomorphology of Coastal Systems

In coastal systems biogeomorphological interactions are clearly demonstrated in shallow, productive waters and various sedimentary environments. Examples of biogeomorphological interrelationships include sand dune development, tidal flats, salt marshes, mangrove systems and coral reefs. Relevant geomorphological factors in coastal systems are bathymetry, bed composition (rock, gravel, sand, silt), and the transport of sediment. It also includes factors that drive morphological processes, such as water flow and wave energy. The biota involved in coastal Biogeomorphology include plants and animals, ranging from very small (microorganisms) to very large (feeding pits of whales).

The geomorphological directly affects biota through its influence on habitats. Coastal morphology and geomorphological processes define the gradients between high and low, between wet and dry and between sedimentation and erosion. These gradients and the processes that cause them determine the gradients in grain size, nutrient levels, organic matter levels and water content. Plants and animals are adapted to specific conditions and will therefore be found and abundant at specific locations (often forming habitats).

The biological influence on geomorphological processes is through the ability of biota to create, maintain or transform their geomorphological surroundings. This is demonstrated by the influence of vegetation and macrofauna on water flow, sediment erosion and deposition, or by the influence of fauna on sediment characteristcs through bioturbation and biostabilization.

In some cases morphological processes are dominant over biological processes and therefore it is necessary for biota to adjust to their environment. In other cases biological processes are dominant (e.g. coral reefs, salt marshes). The most interesting are those cases where there is a mutual interaction that leads to feedback coupling of processes. When looking for these cases, it is important to examine the spatial and temporal scales of the mutually interacting processes.

Biogeomorphology for Hard Substrates

On rocky shores and coral reefs the typical community of organisms modifies the erosion of substrate. Influenced by abiotic factors such as wave energy, splash water, inundation frequency and -period, depth, desiccation and substrate type, a clear zonation of various cyanobacteria, (macro-) algae, fungi, lichens, molluscs, sponges, worms, sea urchins, fish, etc. can be found. Some of these organisms dwell on the surface of the substrate, while others live within the substrate. Their effect on erosion of the substrate is divided into 'biological corrosion' processes that modify the substrate but produce no erosion product, and 'biological abrasion' processes that generate an erosion product. Grazing, burrowing and boring on or in the substrate involves biological abrasion, and is most apparent in coral reef systems.

Worm casts produced by Arenicola marina on sandy/muddy beach increase bed roughness and sediment resuspension.

Biogeomorphology for Soft Substrates

In soft coastal systems, the interrelationships between geomorphological factors and biota mainly apply to benthic fauna and flora. The presence of benthic species is affected by hydraulic and morphologic conditions, such as water depth or elevation on the shore, current velocity, wave action, salinity and grain size. The biota responsible for changes in geomorphology of soft substrates can be divided into two functional groups, namely biostabilizers and biodestabilizers. Biostabilization leads to increased sediment stability and a reduction in erosion potential, whereas biodestabilization leads to reduced sediment stability and an increase in erodibility.

Biostabilization by Plants

On tidal flats, small algae (diatoms) are capable of affecting the geomorphology. These diatoms can form extensive algal mats and excrete extracellular polymeric substances, which is sticky mucus-like substance made of polysaccharides that glues the sediment together and therefore protects the sediment against erosion.

Seagrass is dependent on clear water, it needs sunlight to grow. A seagrass meadow slows down the current velocity near the bed, therefore reducing the resuspension of sand and silt, and this in turn helps to maintain clear water. Their root system also promotes binding of the substrate. In addition, seagrass meadows encourage deposition of suspended sediment, which leads to the supply of organic material and nutrients needed for growth. Seaweeds and salt marsh plants are capable of modifying their physical environments by damping down wave energy and tidal currents. Salt marsh vegetation can also promote sediment deposition resulting in a gradual elevation of the marsh on the upper shore. The higher the marsh gets, the more vegetation can grow and the better the protection against erosion. Other stabilizing effects result from cementation of beach-rock by cyanobacteria and stromatolite formation by algae.

Seagrass meadow.

Upper shore salt marsh with creek system and evidence of benthic diatoms stabilizing mudbanks.

Biostabilization by Animals

Some macrozoobenthos can actively filter sediment particles from the water column and deposit them on the bed. The presence of a mussel bank, for example, will alter the bed in different ways. Mussels at high densities will protect the underlying sediment from erosion by waves and currents (armouring effect). In addition, mussels are filter feeders and actively remove small suspended particles from the water column and any inorganic material (e.g. silt) is rejected as faeces and pseudofaeces. Some of this material will get deposited and incorporated into the bed, causing a shift in the composition to finer sediments and an increase in bed-level. Animal tube fields at high densities can also stabilize the sediment, because there is an accumulation of fine particles and organic matter between the tubes. The tubes may also reduce near-bed flow and create skimming flow above thus creating a stabilizing effect. Mucus production by the community of microorganisms, meiofauna and macrofauna between the tubes may also contribute to sediment stability.

Example of biostabilization by animals.

Biodestabilization

Benthic fauna may destabilize the substrate by their burrowing and surface deposit feeding activity (e.g. bivalves, snails and crustaceans). The constant mixing and recycling of sediment in the top centimeters of the bed, known as bioturbation, results in a change in the vertical particle size profile. Selective uptake of preferred particle sizes and their subsequent defaecation results in sorting and pelletizing of sediments. Together, the burrowing and the constant movement within the substrate, results in the generation of a surface micro-relief that has a higher hydraulic roughness, which increases turbulence and increases the potential for erosion. Bioturbators also interact with biostabilizers, by destabilizing the sediment as a result of grazing on biostabilizers such as microalgae. On the other hand, bioturbators may promote the microphytobenthos growth by organically enriching the sediment via biodeposition. Furthermore, bioturbation affects the sediment water content, porosity and sediment cohesion.

Example of bioturbation.

Mesocosm experiments by researcher show that the effect of bioturbation by bivalves mainly occurs in muddy sediments and is strongest at moderate flow rates, which are insufficient to erode a cohesive sediment bed, but strong enough to suspend sediment particles when the cohesive bonds have been broken by bioturbation. These experiments also show that the degree of bioturbation is mainly related to the metabolic activity of the individual bioturbators. Metabolism is related to body mass, but increasing body mass does not increase metabolism in the same proportion. A large number of small bioturbators thus produce more bioturbation than a smaller number of larger bioturbators with the same total mass. In order to estimate correctly the effect of bioturbation on the sediment balance, models must therefore be able to simulate the spatial distribution of the size classes of benthic animals. Because the density of bioturbators is usually greatest in the intermediate-high part of the mudflat, their activity counteracts further upward growth of the mudflat.

Research has shown that impacts of (de)stabilizing biota on mudflat morphology and bed composition can be significant and can be quantified by adding a limited number of algorithms to existing hydro-morphological models. Researcher show that the influence of (de)stabilizers might extend beyond the mudflat scale to the estuary scale in cases of meso-tidal highly productive systems, such as the Wadden Sea.

Zoning

Examples of biota acting as ecosystem engineers in the intertidal. Key biotas are divided into two functional groups, the bio-stabilisers and bio-destabilisers, with varying spatial distribution along estuarine gradients (both axial and vertical).

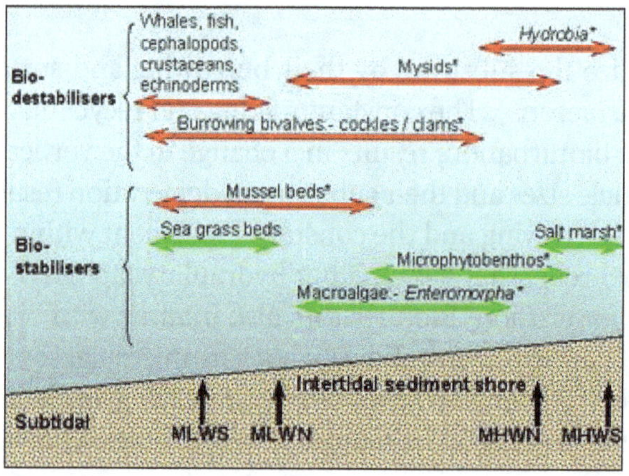

Illustrates how biota act as ecosystem engineers and influence sediment stability and morphology in the intertidal zone.

Urban Geomorphology

Geomorphology has a valuable role to play in the planning, development, and management of urban areas in dry lands, especially in evaluating terrain prior to urban development, and in monitoring changes during and after development. The type of information required depends chiefly on the local geomorphology and the responsibilities and altitudes of the relevant agencies. It will also vary with the phase of regional planning, city planning, site planning and development, and post construction management all require information that can be provided by the geomorphologists.

Much work in geomorphology is of great potential value to man in his use of the physical environment and the application of geomorphologic knowledge has increased in recent years, in harmony with growing public and political awareness of environmental problems and specifically because geomorphologists have come to give readers attention to those aspects of the subject of greatest practical application- the dynamic relations between landforms, materials, and contemporary processes. As far as Applied Urban Geomorphology is concerned it is the study of landforms, and their related processes, materials, and hazards, in ways that are beneficial to planning, development, and management of urbanized areas or areas where urban growth is expected.

Geomorphic knowledge is of great significance in deciding the urban growth and urban morphometry in geomorphological fragile zone. Since urbanization is continues process of urban growth, it is being affected by several factors. An urban agglomeration denotes a continues urban spread and normally consists of a town in its adjoining urban outgrowth. Geomorphic aspect plays a great role in deciding the organizational implications.

One important qualification must be made about urban geomorphology: different aspects of geomorphology have been studied in urban areas for many years- not only by geomorphologists, but also by engineers and others, some of whom may never have heard of the science. It is not, and never has been the exclusive preserve of the geomorphologist. For example, engineers in Los Angeles have studied the movement of sediment into, through and out of the metropolis for decades, and the information they have collected has been used to predict, inter alia, the life span of reservoirs. Nevertheless, as the field of urban geomorphology is one in which many different aspects of the environment are closely related and can be beneficially integrated, and is also one of rapidly growing knowledge requiring greater specialization among those studying it, it is scarcely surprising that planners and engineers are increasingly either receiving geomorphological training themselves or turning to specialist geomorphologists for help in tackling problems of a geomorphological nature.

Recent study suggests that there will be significant variations in the amount of urban expansion most of the urban expansion will occur in China; some specific regions will have high probability of urban expansion; some regions will have low probability of urban growth.

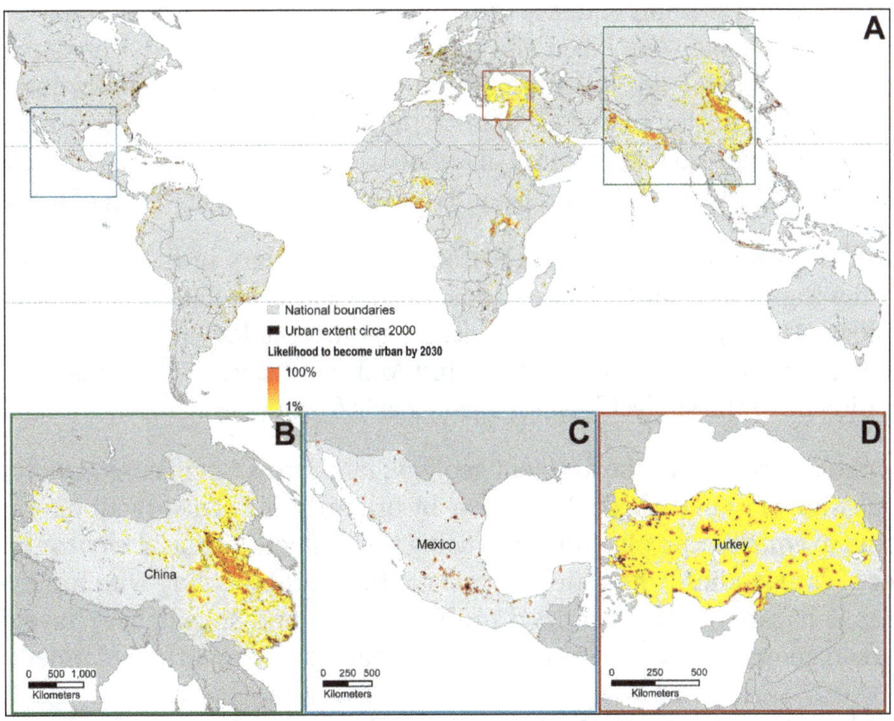

Urban expansion probabilities at global level.

Review in Urban Geomorphology

Wolman was among the first geomorphologists to measure the physical impacts of urbanization on watersheds and stream channels. His studies found that average sediment production rates are moderate to high during pre-urban agricultural uses of land, followed by a spike during construction, and finally a decrease in sediment yield after urbanization. Wolman's results indicate that the channel response following urbanization is a period of deepening followed by lateral migration and channel widening. These observations of the changes in channel form are consistent with other work carried out during the same time period when fluvial geomorphology in the urban environment was a developing scene.

Graf observed the effects of rapid urbanization on the fluvial geomorphology of two small watersheds near Denver, Colorado. Graf found that the initial impact for these sites was extreme aggradation and increased flood plain access due to increased rates of upslope erosion. The secondary impact, after the watershed development was nearly complete and impervious cover in place, was vertical incision and down cutting through the previously aggraded material. Arnold, C.L., P.J. Boison, and P.C. Patton, conducted a similar geomorphic and hydrologic study of a small urbanizing watershed. The frequency of bank full discharge increased, and changes in the sediment regime were consistent with Wolman's observation describing the effect of urbanization. Elsewhere, in other parts of the world, studies in urban geomorphology have gained ground. Notable studies on this line worth mentioning are "geomorpohology and urban development in Manchester area" by Ian Douglas. This work underlined the impact of geomorphology on river dynamics, urban growth, glacial deposits, subsidence, sewer collapse and ground conditions. There are also other works such as urban geomorphology in Dry Lands by Cooke, R.U., Brunsden, D. Doornkamp, J.C. and Jones, D.K.C.

This study was undertaken as a consequence of serious soil erosion, landslides and widespread flooding where hundreds of people were killed and thousands of homes ruined. The dominant environment processes responsible for this crisis are geomorphological problems, problems relating to the nature of land surface and the forces that act upon it. In India the study of urban geomorphology, first appeared in 1988. This study was experienced in Mussoorie and its Environs' by H. Prasad. This study underlined the impact of geomorphology in identifying areas for establishment of new settlements.

Attention to urban geomorphology has increased in recent years in harmony with the growing recognition of the importance of the much broader, but closely related, fields of environmental and urban geology. Although some aspects of urban geomorphology have been considered in recent books, such as those by Coates, Detwyler and Marcus, Legget, Cooke and Doornkamp and Leveson, the rationale of the subject has not been clearly formulated.

Urban geomorphology combines the ambient geology, landforms, and geomorphological processes with the evaluation of impacts brought to these by urbanization. The practitioners of urban geomorphology tend to concentrate on alteration, using the ambient physical environment as a baseline. A number of case studies from different parts of the world (dealing with topics such as slope instability, seismic hazards, increased flood problems, and land subsidence) have demonstrated the utility of urban geomorphology to engineers, city managers, and urban planner.

Objectives of Urban Geomorphology

Geomorphological Appraisal Prior to Urban Development

The urban planner's primary environmental requirement prior to urban development is a knowledge of the nature and disposition of natural resources and hazards. The principal objectives, therefore, of surveys designed to satisfy this requirement are the identification of the range of possible locations of resources and hazards and to analyse conditions within suitable locations, to use the environmental resources more economically, beneficially, and efficiently.

Within these broad objectives, the geomorphologists commonly have several aims:

- To prevent urban growth from destroying valuable resources.
- To identify and evaluate land and material resources required for development.
- To limit undesirable impact of urban development on geomorphological conditions.
- To predict the potential responses of ground surfaces to urban development.
- To assess the potential impact of geomorphological hazards on the urban community.

Geomorphologists commonly adopt one or more of the three major approaches to appraisal prior to urban development. First, by far the most profitable and widely used approach is that of formally classifying and describing terrain features, through morphological or geomorphological mapping and/or the interpretation of air photographs or other remote sensing imagery. Second, analysis of process dynamics and landform change may be accomplished through, for example, the analysis of historical records (e.g. climatic and hydrological data of a region). A third approach is to appraise one, poorly known situation by analogy with another similar but better documented situation elsewhere. This approach is, of course, dependent on the availability of data from the analogous situations and it provides a strong argument for the collection of information in data banks such as that envisaged by the VIGIL network and in deserts by Bekett and others.

Before urban development, the geomorphological contribution is mainly of importance in providing surveys of direct use in themselves, of value as a source of derivative maps etc., and as a basis for more detailed subsequent surveys.

Objectives of geomorphological appraisal during and after urban development During and after urban development the urban planner normally require to know the effects of natural events and circumstances on the urban community, and the impact of urban development on the environment. It is to be noted that the primary interest of the planner is to understand the environmental consequences of urban growth. Within this primary objective, planning and management aims include:

- The minimization of environmental impact.
- The development of local, spatial and temporal data bases from monitoring studies to formulate the urban development plans.
- The continues evaluation of plans, management organizations, and procedures to ensure an harmonius environmental management.

The main aim of geomorphological work in this context is to monitor the dynamics of geomorphological systems with a view to predicting spatial and temporal changes in a way that allows the planner to respond effectively and in good time. Geomorphological approaches to appraisal during and after urban development are similar to those adopted prior to development, but their relative importance changes. Field monitoring is pre-eminent, whether it is on a global scale or a city wide scale. Monitoring often requires the establishment of fixed observation stations.

Examples of the Value of Geomorphological Information in the Planning Process

1. Geomorphology in city planning: If, in the planning of a new city the geomorphological surveys are carried out before the settlement of the city, then we can avoid incongruity between the environmental conditions and the city. For example, suppose urban planners making the city plan of a city, the first city plan (we can term it is as city plan A) has an attractive spatial geometry, but it would have encountered several problems- building over scarce aggregates; hydro-compaction; flooding, sedimentation and erosion or roads crossing alluvial-fan channels; blowing sand and salt weathering. In the revised city plan (we can term it as city plan B), which attempts to accommodate the implications of the geomorphological map, most of these problems are either avoided or sensibly controlled, thus saving resources, time and money.

At the scale of a whole city, a common problem is that the management of a single, natural unit, such as a drainage basin, is divided between several administrative organizations, with resulting duplication or dispersion of effort, and perhaps competition and conflict. This is a problem that can be avoided if environmental criteria are used intelligently in the initial formulation of responsibilities of different authorities- although it would be naïve to assume that other factors are not usually more important in such formulations.

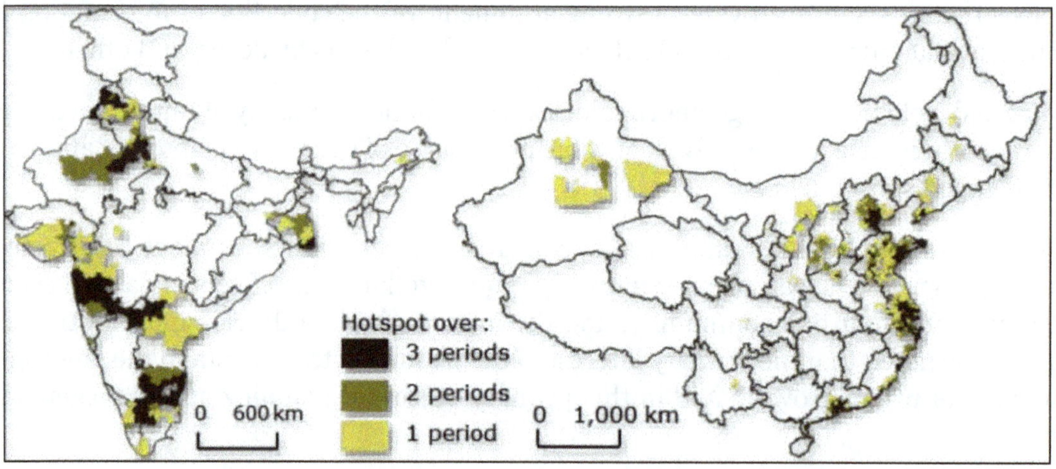

Urban cluster growth hotspots detected from DMSP/OLS nighttime lights in China (at county level) and India (at district level) over three periods: 1992-1997, 1997-2003, 2003-2010.

2. Geomorphology and site planning and development: An excellent example of the way in which geomorphological advice can beneficially modify site development plans is provided by Mader and Crowder. He cited the examples from USA's hilly country areas. In 1956 a residential development was proposed in the hill country of the growing settlement of Portola Valley, south of San Francisco, California. Subsequent geological and geomorphological studies related to the formulation of the town's general plan and zoning and subdivision ordinances revealed the nature and extent of potential slope instability in the proposed development area. A relative slope stability map showed areas of stable, potentially moving, and moving ground, and located major landslides. That map, together with other relevant information formed the basis for a new plan, in which houses were

clustered on stable ridge crests, and the number of lots was only slightly lower than the maximum permitted under the general formula relating lot size to average slope; even more important, about 15 houses sites and some roads on the original plan were removed from actively moving ground, and considerably more house sites were removed from potentially moving ground. There are numerous instances where specific geomorphological information might have helped to avoid problems and made the site-development process more efficient.

3. Geomorphology and post-construction management: It is suggested that major environmental hazards can be avoided by monitoring of geomorphological processes and landform changes after construction and this can also help environmental managers and policy makers in influencing future policy planning. Ruby's study of sediment-yield trends in the Los Angeles River catchment is a straightforward example. The accumulation of sediment in debris basins at the mouths of mountain canyons provides a rough measures of sediment yields and allows the performance of smaller check dams to be evaluated. Ruby compared accumulated sediment yields in one canyon before and after check-dam construction with the regional norm. There are many problems associated with the data and their interpretation, but regression lines relating sediment yield in Dunsmore canyon to the regional norm for 30 years of records show that for the first 21 years the canyon performed similarly to the regional watershed, that during the ten years following construction of check dams sediment yield from the canyon relatively declined, and that the decline has become progressively less over time. In the period since check dam construction sediment yield was reduced overall by approximately half, but there now appears to be a trend back to pre-treatment yields, indicating a decline in the trap efficiency of the check dams. Clearly an alternative strategy for sediment control is required.

Before geomorphological problems of any urban development are reviewed systematically, two further introductory themes require examination: the availability of information relevant to geomorphological studies in urban development area, and the integration of geomorphological data into the whole assemblage of environmental data relevant to planning decisions.

Relations between Geomorphology and other Scientific Information of Value to Planners of Urban Areas

Geomorphological information forms but one part of the body of environmental data that may be of value to urban planners and engineers. Many environmental attributes of interest are both closely linked and highly interdependent. Data on the regional setting is primarily relevant in assessing hazards and resources and in choosing locations for development; data on site conditions related to decisions about specific developments. On the basis of recent studies it can be argued that geomorphological surveys can provide a useful first stage in the environmental assessment, not only because geomorphology is a fundamental basis for urban development, but also because studies of geology, soils, hydrology, etc., can all benefit from, and be facilitated by access to geomorphological surveys.

Geomorphology and Environmental Impact Assessment

On the basis of above discussion it is important to record that demands are increasing from planners for wide-ranging environmental reports prior to making and implementing planning decisions. The most important new requirement is not so much for resource survey but for studies to evaluate the impact a proposed development is likely to have on the physical environment.

Although many planning authorities have required environmental impact assessment for years, recent legislations has enforced and codified the requirement in several countries, providing a substantial stimulus to the development of methods for assessment and for integrating environmental data. The most important new law was NEPA (the National Environmental Policy Act), 1969, passed by the US government, and this was followed by similar measures in other countries such as Australia and Israel.

Climatic Geomorphology

Climatic geomorphology is the branch of geomorphology which deals with the effects of climate on geomorphological processes and consequently on the character of land-forms. It has included the identification of climatically controlled zones and has attempted to define provinces with distinctive denudational processes.

Hydrogeomorphology

Resources are essential elements for the sustainability of human beings. Among all the resources, water and land are among the most important resources. Hydrogeomorphology deals with both of these resources. Hydrogeomorphology is a new field in geosciences that has emerged to study the linkages between water processes and landforms. It is a combination of three different terms:

- Hydro: Refers to water including both surface and underground water.
- Geo: Refers to the ground and the landforms.
- Morphology: Refers to the surface characteristics of the landforms.

Hydrogeomorphology studies to major aspects:

- Affects of moving water to hill slopes, rivers and landscapes on geomorphic processes and geomorphic forms.
- Affects of geomorphic form on the spatial and temporal distribution of shallow groundwater.

So hydrogeomorphology is an inter-disciplinary science that studies the link between two important branches of geosciences i.e. hydrology and geomorphology. Other than hydrology and geomorphology, the concepts of hydrogeomorphology are also largely derived from the different branches of geosciences like geology, remote sensing, climatology and natural hazards.

To study hydrogeomorphology, it is important to understand the difference between three branches of geosciences i.e. hydrogeomorphology, geology, hydrology and Geomorphology Hydrogeomorphology, geology, hydrology and geomorphology all are different branches of geosciences and covers different subject matter but often get confused. Geology is that branch of geosciences that studies materials of the earth along with the processes that are working upon them. Hydrology studies the occurrence, distribution, movement and relationship of water with all the aspects of environment it interacts with. Geomorphology is the subfield of geology and studies the structure and evolution of the surface of earth. It deals with origin and changing structure of earth's landforms.

Fundamentals of Hydrogeomorphology

The important fundamentals of hydrogeomorphological studies are given below:

1. Mechanism and Process: Mechanism gives the explanation by describing the physical and chemical effects while the process is the simultaneous operation of a set of specific mechanisms over a specific period of time. In hydrogeomorphological studies the landforms are studied with reference to groundwater conditions of the area. So while studying the mechanism and process in Hydrogeomorphology, the morphological, climatic and hydrological criteria's are considered. The study of mechanism and the process involved helps to give explanation behind morphology and distribution of landforms.

2. Basic tools required: To carry out a study in hydrogeomorphology it is very important to first design a geographic database including both spatial and non-spatial data requirements with their sources. The most important requirement for the studies in hydrogeomorphology is maps. This includes the topographical maps, geological maps of the area, soil map, rainfall and climate distribution map, geomorphological map, population density map and groundwater fluctuation map.

3. Properties of Earth Materials: Hydrogeomorpholgists studies the properties of earth materials to get better understanding of mechanism and process behind the geomorphic features. They emphasis on the properties of the parent material influencing the formation and development of landforms along with the hydrological conditions. Type of rock, weathered material, soil, superficial deposits, shear strength, porosity and mineral composition are some of the important properties that hydrogeomorphologists studies during their research.

4. Spatial Scales of Hydrology: Hydrogeomorphology not only studies the effect of hydrological processes on the geological processes but also analyses the effect of landforms on the hydrology of an area. Thus the study of scaling effects becomes essential. The spatial scales in hydrology can be distinguished into three types:

- Local Scale: On local scale the parameters like slope angle directly influences the water flow path geometries, flow velocity and quantity.
- Hill Slope Scale: In hill slope scale is characterised by its runoff production. The soil properties influence the runoff production in the hill slope scale.
- Catchment Scale: In catchment scale morphometry of the basin influences the runoff production.

5. Dimensions of Hydrological Units: Hydrogeomorphology studies the role of dimension of hydrological units in defining the size and dimensions of the geological landforms. Watersheds are delineated at various levels on the basis of the drainage network. Watershed can be defined as an area of land that includes common set of streams and river draining their water into single larger water body. Water in watershed can come from any of the sources of water from precipitation, in form of rain or snow or as groundwater or surface runoff. The size of the watershed depends on size of the stream, river, the point of interception of stream or river, the drainage density and its distribution.

Hydrological Cycle and Water Budget

Hydrology being an important part of hydrogeomorphology it becomes important for hydrogeomorpholgist to study the hydrological cycle, water balance and water budget for their research studies. Circulation of water into different forms between various spheres of earth surface is called hydrological cycle. Hydrological cycle recycles earth's water. The hydrological cycle includes inflow, outflow and storage of water at different levels. The inflows add the water into the system while outflow subtracts the water. Storage helps in the retention of water in the system.

Water budget studies the water availability. It includes the balance between the inflows and outflows of the water. In water or hydrologic budget the inputs are derived from precipitation, surface water inflow and groundwater inflow. Output factors include evaporation, combined surface and groundwater outflow and transpiration.

Application of Remote Sensing and Geographic Information Systems in Hydrogeomorphology

To get the proper understanding of hydrogeomorphology, it is essential to collect geological, structural and hydrological data of region. Collection of hydrological data i.e. assessing the data of surface and sub-surface water resources requires huge time and manpower. GIS and remote sensing are the platforms through which the all the required data can be collected with better accuracy and time. Remote Sensing and GIS carries great importance in hydrogeomorphological studies. Before looking at the application of remote sensing and Geographic Information System (GIS) in hydrogeomorphology it is important to understand the concept of remote sensing and GIS.

In early 1960s the term remote sensing was first used for the first time. Remote sensing can be defined as the process of acquiring information through recording devices called sensors about the objects and phenomena without getting in physical contact with them. In remote sensing the data acquisition is done through different stages. The stages involved in remote sensing for data collection are given below:

- Source of energy.
- Transmission of energy from the source to the earth surface.
- Interaction of energy with earth's surface features.
- Propagation of reflected energy/emitted energy through atmospheres.
- Detection of reflected/emitted energy by the sensor.
- Conversion of energy recorded into digital/photographic form.
- Extraction of the information from the data generated.
- Conversion of information into map/tabular form.

Satellite imageries and aerial photographs are the outputs of the process of remote sensing passing all the above mentioned stages. The interpretation of the satellite images and aerial photographs generated by remote sensing are made through the different elements of visual interpretations. The important elements are tone, texture, size, shape, shadow, pattern and association. In the study of hydrogeomorphology the same elements of visual interpretation are used to get the information through the interpretation of the images generated by remote sensing of the landforms and the regions covering the water resources.

Geographic Information System (GIS) is a tool of digitally capturing, managing, analyzing and displaying geographic data using the computer hardware and software. Maps, computer hardware and software, information, procedures and people are the important components of GIS. GIS is an important tool for geographic studies because it analysis and answers real world problems. It makes dynamic maps and displays detailed information about the features in the maps. It not only displays but also studies and establish relationship between the features. The application of GIS can be seen in studying the relationships, patterns and trends of various spatial and non spatial elements of the earth surface. GIS contributes vastly in the study of Land use/land cover studies along with the studies associated with water and soil resources.

Over the years due to rapid increase in population, rapid urbanisation and industrialisation along with failure in monsoons has restricted the availability of surface water. Limited availability of surface water has increased the burden on groundwater resources. This has resulted into higher rates of groundwater withdrawal which further resulting in depletion of groundwater at alarming rates. For the management and conservation the study of groundwater resources is very important. Satellite remote sensing has made it easy to study the spatial distribution of ground water prospects on the bases of geomorphology and other associated features. Satellite remote sensing is also a useful technique for groundwater exploration along with delineating the hydrogeomorphological units. Ground water accumulation, infiltration and movement largely depend on the factors like drainage, geomorphology, and slope of the terrain, vegetation, soil and depth of weathering. All these factors can be easily studied using remote sensing at various levels.

Remote sensing not only studies the hydrological aspects but it also an effective tool for geological, structural, geomorphologic analysis and their mapping due to its synoptic, multi spectral and multi-temporal capabilities. Geologists largely depend on satellite imageries to collect the data on various lithological units. The important advantages of remote sensing in hydrogeomorphological studies are as follows:

- Remote sensing has access to large areas and even in inaccessible areas.
- The aerial photographs and satellite imageries provide detailed information about the uppermost layer of the earth surface which is essential for the hydrogeomorphological studies.
- Digital enhancement of satellite imageries improves the level of information useful in study of hydrogeomorphology.
- Data generated by remote sensing provide more accurate and spatial information in comparison to the hydrogeological surveys.
- Through remote sensing hydrogeomorphological mapping of a terrain and analysis of their processes can be done easily.

- This will further help in soil resource mapping, groundwater potential zones demarcation, landscape ecological planning, hazard mapping and their environmental applications.
- The application of geomorphologic mapping using remote sensing can also be seen in land use planning and water resource management.

The IRS-1C and 1D data of WiFS, LISS-III and PAN sensors are highly useful for geological mapping. The WiFS camera gives the synoptic coverage of large areas and thus is useful for the regional scale mapping and understanding. The finer geological features like the traces of bedding and minor joints can be easily are identified through panchromatic data. The panchromatic data provides detailed mapping while the multispectral LISS-III gives semi detailed mapping.

Along with remote sensing, in recent years several techniques have been developed in the field of Geographic Information System (GIS) through which hydrogeomorphologic studies can be conducted. In recent years GIS has emerged as a powerful tool in analysing the various aspects of groundwater occurrence. Groundwater resource of any area is largely controlled by the factors like lithology, structure, geomorphology, slope, drainage and land use/land pattern. All these factors can be studied and analyzed as thematic layers using Geographic Information System (GIS). This makes it easy to delineate the groundwater prospect and deficit zones. The main advantages of GIS in hydrogeomorphological studies are as follows:

- Large volume of data can analysed and integrated using GIS.
- Manipulates and analysis the individual layer of spatial data.
- Rapid, accurate and cost effective tool.
- GIS is a powerful tool for the generation of hydrogeomorphological mapping.

Application of Hydrogeomorphology

The application of hydrogeomorphology can be seen in planning and management of various activities on the earth surface. Some of the important applications of hydrogeomorphology are as follows:

- The accurate, detailed, timely and reliable data on the extent, location and quality of land along with water resources and climatic characteristics helps resource planners in agricultural landuse.
- Data on land potential and conservation requirements through the hydrogeomorphological studies helps in improving the quality of land.
- Hydrogeomorphological studies are found helpful by environmentalists in identifying hazards and studying climate change.
- Geologist have found it useful in examining role of surface and subsurface flow regimes and flow paths on fluvial erosion and mass wasting.
- Ecologists found it useful to describe the linked water and geomorphic conditions that define habitats in wetlands, rivers and other environment.

- Watershed management depend on the collection and management of information on physical relationship between vegetation, soil and water resources, which can easily be done by hydrogeomorphological studies.

Future of Hydrogeomorphology as Geosciences

Over the years the relationship between physical and human environment is taking new turns and shape. The need for the new subfield called hydrogeomorphology stems from the emergence of various challenges resulted from growing human population combined with its effects on environmental and water resource systems. Transformation in hydrological cycle, water use, land use and climate is result of this change in man - environment relationship. In today's era hydrogeomorphological studies are great option to manage and control the environmental problems. In recent years hydrogeomorphology has become an important branch of geosciences to tackle natural hazard impact, environmental auditing, resource assessment and impact assessment. To deal with the new challenges, hydrogeomorphology has developed close relationship with various fields like ecology, soil science and pedogeomorphology.

The growth in the importance of hydrogeomorpholgy as a geosciences can be seen with the fact that now the world organisations like UNDP and World Bank have added a clause of understanding the hydrogeomorpholgy the area before starting any development project into it. Emphasis has been made by these organisations to have detailed knowledge of landforms, hydrogeology materials and earth surface processes to be utilized or the remedial work, planning framework and land zoning plans. Thus it can be said that the hydrogeomorphological knowledge is now being utilized in planning and development of earth as a whole. This shows the growth in status and responsibility of hydrogeomorpholgy in recent years as an important field of geosciences.

Soil Geomorphology

Soil geomorphology addresses geomorphic details and dynamics at and near the Earth's surface that affect, or are affected by, soil processes and products. It specifically addresses the distribution, properties, and dynamic behavior of soils—issues that traditional geology and geomorphology do not emphasize because of scale or minimize because of perspective. These soil issues are particularly environmentally and economically meaningful because they occur at the "human-operative scale," i.e., the scale at which most land use decisions are made and applied and their consequences felt.

Soil science, particularly soil geomorphology, is based upon a robust relationship between lithology, hydrology, stratigraphy, geomorphology, and, to a slightly lesser degree, biota and climate. In many settings, hydrology and hydropedology dominate the physical redistribution of materials and catalyze the chemical reactions that transform earth materials.

Boundaries and Transitions

Some parts of the landscape and the soils on them are separated from their neighbors by distinct, sharp boundaries over a lateral distance of just meters. For example, a stream terrace may be sharply separated from adjacent cliffs and talus cones by an abrupt, easily observed scarp. Other parts of the landscape and the soils on them have lateral boundaries that are very gradual and indistinct, transitioning over tens of meters or kilometers. For example, a loess mantle thins gradually with increasing distance from the source of the loess.

Distinct, sharp breaks between landforms are evident over short lateral distances, as shown by the talus cone in this canyon along the Palouse River in Washington.

Scale

Another major determinant in conveying soil and geomorphic information is the scale of interest (e.g., local, regional, global), which is established in initial survey perspectives and goals. Scale partly predetermines what is "relevant" and what is not, what must be "shown" and what cannot. A person managing a 1-hectare home site has different information needs and relevancies than a regional planner who evaluates and manages a city or State. The scale of interest can be quite different for different users, and the appropriate level of information to be collected and delivered differs with the scale. At regional scales, it is typically appropriate to evaluate and emphasize landscapes and very large, constituent landforms and to minimize smaller landforms and micro-features. For example, in an order 3 reconnaissance survey, dune fields, mountain ranges, or bolsons may be evaluated. Conversely, for localized surveys, it is typically necessary to focus primarily on smaller landforms, micro features, or pieces of landforms and to give only nominal attention to landscapes. For example, in an order 1 survey, barchan dunes, slip faces, or head slopes may be evaluated.

Digital Soil Mapping and Scale

Widely available GIS tools and digital soil mapping methods have made it much easier to move between scales. For example, geospatial information that was previously constrained by scale may be combined or split apart. It is now possible to produce: (a) primary surveys capable of spanning a much wider range of scale, and (b) derivative or second-generation resource maps from a primary spatial database that span wider ranges of scale than the initial survey. The capability to move between scales must coincide with primary data that accommodates such changes in scale. For example, traditional soil inventories were made based upon stated, relatively narrow spatial scales (e.g., order 2 or 3). The initial field data collected, assigned classes, spatial separations (e.g., polygons), interpretations, and other information remain constrained by the rules and decisions for the scale at which they were originally developed. It is typically possible to aggregate data upwards in scale with a minimal impact on information integrity. It is much more challenging to do the opposite and legitimately disaggregate primary data to create finer resolution information. With new tools, software, and statistical methods, it is possible to produce information with finer resolution using disaggregation techniques. However, the original data, metadata, and inherent decisions made at the original inventory scale remain determinate; those original biases will persist across scales. Some morphometric parameters, such as slope, can be readily replaced or enhanced with higher resolution information, such as LiDAR. In some cases, such as the separation of soil bodies at finer resolution, previous information can be more problematic and new supporting data appropriate to the new scale may be needed.

Capturing Soil-Landscape Relationships at Various Scales

It is typically impractical or financially prohibitive to make an ideal number of field observations. Consequently, representative observation sites need to be chosen wisely and used to extrapolate to areas that cannot be visited. Observation efficiency can be greatly enhanced by developing soil-landscape models that capture recurring spatial relationships. Part of such model development includes selecting an appropriate scale. Scale determines what can and cannot be shown. Robust soil-landscape models are subject to incremental changes and refinements that reflect the accumulation of knowledge and data during the progression of the survey.

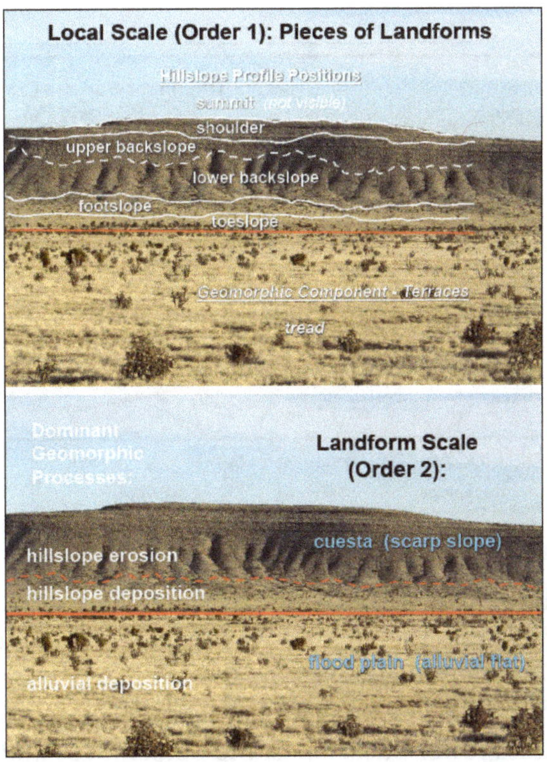

Figure shows a scarp slope of a cuesta above an alluvial flat. Scale determines which geomorphic descriptors can be effectively used and presented. Geomorphic evaluations of regional scope (landscape scale) can separate tectonic hills from areas dominated by fluvial processes (not shown). The more localized, landform scale can differentiate dominant landforms (cuesta and alluvial flat) and the dominant geomorphic processes that control them within the landscape (top image). More detailed erosional and depositional surfaces derived from dominant hillslope processes need to be represented at a finer scale. These include microfeatures (e.g., ribs and groves), hillslope profile positions (e.g., summits and shoulders), and geomorphic components (e.g., components of terraces or hills, such as nose slopes, head slopes, and side slopes; not shown).

A site may have one of three common scales of models: a large area or catchment (landscape) scale, a hill (landform) scale, and a hillslope position or pedon (microfeature) scale. The appropriateness of each scale depends upon the perspectives and objectives of the survey. Each scale can be applied, but each conveys somewhat different information and has different strengths and limitations. For example, for a setting along the border of the Gypsum Plains State Physiographic Area in Culberson County, Texas, a series of hierarchical (nested) landscape models can be developed and applied. At a landscape scale, the area presents eroded structural hills and alluvial plains. The dominant geomorphic processes are tectonic and erosional (and, to a minor extent, fluvial) and are regional in scope. At a landform scale, which is finer than landscape, the same area presents an eroded, structural hill (questa) and valley floor (alluvial flat). The dominant geomorphic processes are erosional and fluvial and are local in scope. At a microfeature scale, the area presents discrete subsets within landforms. The dominant geomorphic processes are hillslope erosion and deposition (slope wash processes) and fluvial modification (alluvial deposition processes) and are very localized in scope. Different scales yield different information.

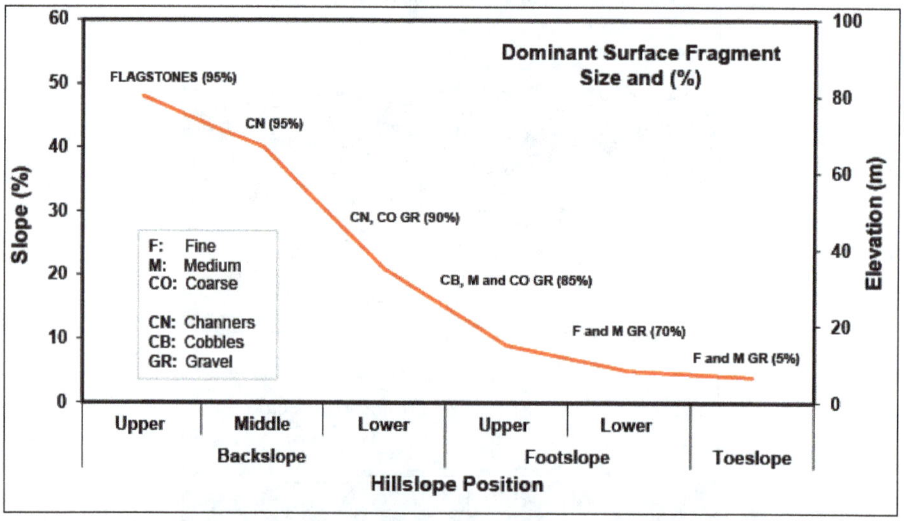

Figure shows diagram of the change in surface fragments along a transect of a scarp slope on a cuesta in Culbertson County, Texas. The progressive reduction in dominant size and percentage as one moves down slope demonstrates the impact of hillslope processes at a local level. On the upper backslope, dominant erosional hillslope processes are evidenced by the presence of flagstones on the surface. On the middle and lower backslopes, the percentages and sizes of surface fragments (channers and others) are indicative of lateral transport. On the lower footslope and toeslope, rock fragments are dominantly medium and fine gravel with decreasing percentages. They are indicative of deposition processes.

In the digital environment, grid scales can be used to determine which resolution best captures important attributes for landscape modeling. What is apparent at one scale may not be apparent at another. Digital smoothing techniques, while appropriate for some analyses, can adversely affect the ability to identify features at a given scale.

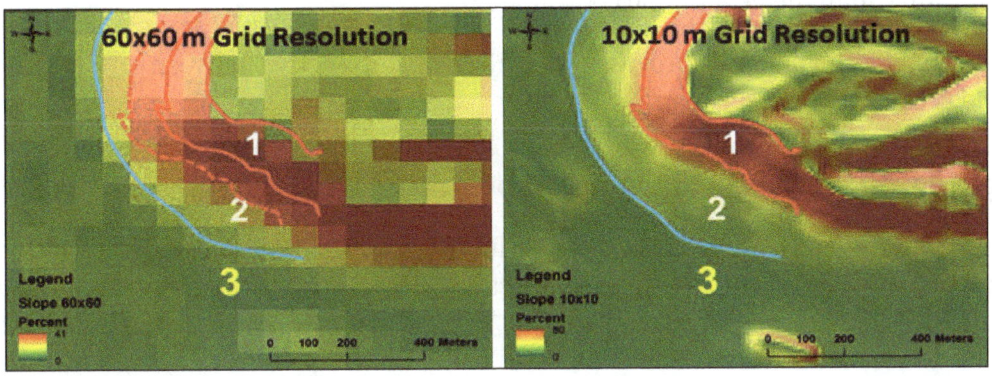

Figure shows a comparison of digital maps with 60 × 60 m (left) and 10 × 10 m (right) grid sizes demonstrates the detail at different resolutions. The 60 × 60 m grid visually captures the three major landform units but not the geomorphic components of hills, namely nose slope, side slope, and head slope. An even finer resolution, such as LiDAR elevation data, is needed to capture the physical expressions of very localized hillslope processes.

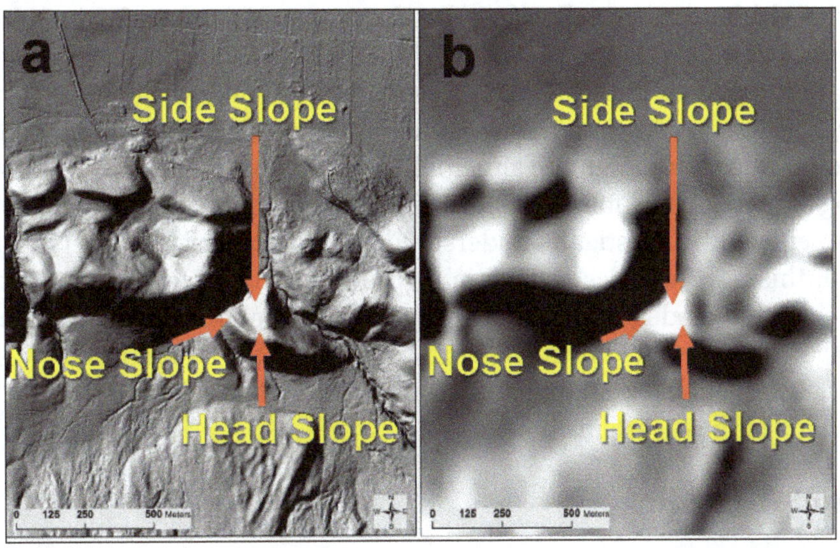

Figure shows images showing changes in slope class interpretation as affected by digital elevation model (DEM) resolution from LiDAR. Image a—Grid size (1 × 1 m) captures surface features nested within hillslope position. Image b—As resolution decreases (30 × 30 m)and surface becomes digitally smoother, the geomorphic components of hills (nose,side, and head slopes) are no longer visible.

Placing Soil-Landscape Relationships

Soil geomorphology can be used effectively to evaluate, recognize, and communicate the context (setting) of natural systems. Context is key to:

- Recognizing and understanding the materials and processes occurring at a given location or area.

- Effectively predicting the distribution of materials that dominate the area. Commonly, the first and most useful question is "Where are you on the planet?" (i.e., the geomorphic context rather than the geographic coordinates).
- Recognizing the dynamic processes and relationships active between and within soils and land areas.

Development Stages for Soil-Landscape Models

Developing accurate landscape models begins with using expert knowledge and experience to anticipate and portray relationships between landscapes, geomorphic systems, and soils. In whatever form, an initial model must be rigorously tested ("ground-truthed") and revised as needed to accurately reflect actual, observed relationships of soils and landscapes.

Developing Initial Soil-Landscape Models

It is helpful for scientists to have a general expectation of what they will encounter in an area or at a given site. For example, if the site is in a river valley, it is reasonable to expect that the site will reflect fluvial dynamics and features in the landforms, sediments, and soils of the area. The location of the site within the river valley (e.g., headwaters vs. delta) can further refine preliminary expectations. Headwater fluvial sediments, and the soils that formed in them, may be expected to contain more, larger, and less rounded coarse fragments, be somewhat less well sorted, and have less contrasting sediment strata than fluvial sediments in a delta. In addition, general conditions and dynamics for a water table might be anticipated (e.g., gaining stream vs. losing stream) based on the prevailing climate. A preliminary model can provide a tentative framework in which scientists can begin to efficiently investigate, understand, and give order to a diverse natural world. In the digital world, this now includes considerable pre-mapping, based primarily upon improved, detailed digital elevation models and existing information. Preliminary soil-landscape models are helpful. The adage "If we hadn't known about it, we wouldn't have recognized it" is a relevant and practical truth. The more one understands a natural system, the easier it becomes to recognize the physical expressions of those systems.

Describing and Recording

Although soil-landscape models can be very useful at any stage of soil inventory, it is critical that a scientist maintain an open mind and adjust models as new information becomes available. A common error by field workers is to continue to defer to a preconceived expectation, such as a preliminary landscape model, rather than to adapt their ideas based on what is actually encountered. If a landscape model fails to match the natural features and sediments actually found, it must be modified or abandoned. A useful model should fit the facts, not the reverse.

Consistently Describing Landscapes, Landforms and Geomorphology

There are various major kinds of information typically gathered to identify, evaluate, or communicate geomorphic information. A descriptive system and set of terms that can be used consistently are important for conveying the information accurately between individuals and making valid comparisons from place to place. The NCSS generally uses the Geomorphic Description System, or GDS. The system consists of three main sections:

- Physiographic location: A named geographic area is specified with a defined location.
- Geomorphic description: A discrete land surface feature (a separate entity) or an assemblage of features is identified. Features are categorized by dominant process of origin or geomorphic setting.
- Surface morphometry: Land surface shape or geometry is described. A discrete portion of a geomorphically defined land feature, area, or slope segment is identified. Microrelief, drainage patterns, and other surface features are also described.

Geographic and Physiographic Information

Geographic and physiographic information primarily addresses the question "Where is it?" It identifies the specific location of an area on the planet (e.g., the Appalachian Mountains). This information has powerful communication value to a wide range of people. Most people can relate a geographic name to some understanding of an area. Fewer users appreciate explicit technical descriptions (e.g., an eroded, folded, paleo-continental margin mountain system). The Appalachian Mountains and the Rocky Mountains are both mountain systems in the geomorphic sense, yet they differ in important geographic locations as well as geomorphic details. Their names emphasize geographic differences, not the geologic composition or type of mountains. Geographic and physiographic names may or may not accurately reflect the actual geomorphology. Confusion can arise if geographic names include geomorphic terms that do not adhere to their technical geomorphic meaning. For example, an area geographically identified as Thompson's Bench may geomorphically be something other than a true structural bench, such as a stream terrace or a horst.

Physiographic information combines geographic information with limited geomorphic information to describe location. In physiography, distinctions in topography, bedrock or parent material, watersheds, and other attributes are used to group geomorphically similar or related areas. Experience has consistently demonstrated that physiographic information is the most robust, versatile, and least ephemeral basis for describing and partitioning the landscape mosaic. Other location approaches emphasize spatial distinctions based on something other than physiography. For example, some approaches emphasize natural ecological environments or land management systems. The most appropriate descriptive framework for location is determined by the survey perspectives and goals.

Physiographic Location

Because there are numerous and diverse users of soil survey information, there are various perspectives and objectives. A single scale of information cannot best serve all of them. A homeowner has different needs and interests than a regional planner or a resource management government agency with national scope. Information on physiographic location can be usefully partitioned within a generalized hierarchy of scale. In the United States, the highest three levels are based on work by Fenneman for the conterminous U.S. and by Wahrhaftig for Alaska. Additional lower levels describe physiographic areas within States and more localized areas.

Table: Physiographic Location, Relative Scale (in Descending Order) and Examples in the U.S.

Physiographic location level	Relative scale	Example
Physiographic division	Continental scale	Interior Plains
Physiographic province	Regional scale	Central Lowland
Physiographic section	Sub-regional scale	Wisconsin Driftless Section
State physiographic area	State scale	Wisconsin Dells
Local physiographic name	Local scale	Blackhawk Island

Maps are available for the three upper physiographic levels from Fenneman. They are reproduced in the Geomorphic Description System. Maps of State physiographic areas are generally available through State Geological Survey offices or from the local university NCSS cooperators. Information on local physiographic names is primarily obtained from U.S. Geological Survey topographic quadrangle maps, where available.

Geomorphic Description

Geomorphic description attempts to answer the questions "What is it?" and "How did it occur?" for natural features and sediments. Earth surface features can be effectively arrayed in various ways for particular needs. Two ways that complement one another are:

- Master lists loosely stratified by scale into landscapes, landforms, and microfeatures.
- Lists arrayed by geomorphic environment (fluvial, eolian, etc).

Scale

Earth surface features can be partitioned into any number of scale ranges, but three general levels have proven consistently effective: landscapes, landforms, and microfeatures. Landscapes, features at the coarsest scale, are collective groups or families of related landforms and typically cover large areas. Examples are a mountain range and canyonlands. They are most important to general or reconnaissance surveys.

A canyonlands landscape in the San Rafael Swell, Utah.

Landforms are discrete, individual features that are related to one another within the context of the larger landscape and can be mapped at conventional mapping scales, such as order 2. They are typically local in size, but some can be quite large. It is helpful to remember that natural and anthropogenic landforms and microfeatures can be expressed as the result primarily of removal, transport, or deposition. For example, blowouts and borrow pits are the result of removal, longshore bars and active dunes are the result of transport, and alluvial fans and dredge spoils are the result of deposition.

Loess hill and river valley landforms in western Iowa, along the Missouri River.

Microfeatures are discrete, individual, earth surface features that are readily identifiable on the ground but are too small or intricate to display or capture at conventional mapping scales. Examples are vernal pools and turf hummocks. Where present, these mini-landforms can have substantial impact on internal water flow, soil development, natural ecosystems, and land management.

Turf hummock microfeatures in a wet meadow in Oregon.

There are many choices within each of these major categories, particularly landforms. The choices within each category are commonly arrayed as alphabetized master lists, which are particularly appropriate for databases. Because of the huge number of choices, it is helpful to use subsets for the three main categories, arranged by geomorphic environment or other groupings of commonality.

Geomorphic Environment

A geomorphic environment is a natural setting dominated by a geomorphic process of formation and modification and the resultant behavioral dynamics. For example, a fluvial geomorphic environment consists of landforms and associated sediments created directly by, or in response to, channel water flow (fluvial processes). In such a setting, present-day environmental dynamics, such as ground-water and water table dynamics, are likely to be largely controlled by the fluvial system that formed the area's landscape. Table lists the prominent geomorphic environments that are most relevant to soils in the United States and extensive elsewhere in the world.

Table: Prominent Geomorphic Environments and Processes in the U.S. and Examples.

Geomorphic environment or other setting	Dominant process or attributes	Example LS = Landscape LF = Landform Micro = Microfeature
Coastal marine & estuarine	Wave or tidal control; areas near shore, shallow submarine areas.	LS: Coastal plain LF: Nearshore zone Micro: Shoreline
Lacustrine	Related to inland water bodies.	LS: Lake plain LF: Lakebed Micro: Strandline
Fluvial	Concentrated channel flow.	LS: River valley LF: Stream terrace Micro: Bar
Solution	Dissolution and subsurface drainage.	LS: Cockpit karst LF: Sinkhole Micro: Solution corridor
Eolian	Wind related (erosional and depositional).	LS: Dune field LF: Barchan dune Micro: Slip face
Glacial	Directly related to glaciers (glaciofluvial, etc).	LS: Till plain LF: Ground moraine Micro: Tarn
Periglacial	Non-glacial, cold climate (modern or relict).	LS: Thermokarst LF: Patterned ground Micro: Stripe
Mass movement	Gravity	LS: Breaklands LF: Landslide Micro: Sag
Volcanic & hydrothermal	Volcanic and/or hydrothermal processes.	LS: Volcanic field LF: Lava flow Micro: Tumulus
Tectonic & structural	Regional and local tectonic processes or crustal movement.	LS: Mountain range LF: Graben Micro: Sand boil
Erosional	Dominated by hillslope and sheet-wash processes (non-concentrated channel flow).	LS: Breaklands LF: Pediment Micro: Gully
Wetlands	Vegetated and/or shallow wet areas and wet soils.	LS: Everglades LF: Mangrove swamp Micro: Vernal pool
Water bodies	Surface water features; primarily open water.	LS: Ocean LF: Oxbow lake Micro: Pond
Subaqueous features	Permanently submerged features that support plant life or adjacent areas.	LS: Lagoon LF: Lagoon bottom Micro: Shoal

Multiple Geomorphic Processes

Sites may have evidence of more than one geomorphic process. Typically, these processes are not equal. One process tends to dominate a given land area, and other processes, if present, have a minor presence and influence. It is important to recognize the most pervasive process and its products because they establish the primary landscape configuration and sediment composition. It is also important to recognize secondary processes and their sediments where they substantively affect soils. For example, loess commonly mantles till plains in continental glaciated environments. If the loess cap is relatively thin, the dominant and most important geomorphic context is glacial because it explains the prevailing land features, topography, and unconsolidated materials. The loess should be recognized for what it is, and not some other geologic deposit, and correctly identified and described in the soil stratigraphy. If the loess cap is thick, it can supersede underlying glacial materials and function as the determinant geomorphic setting itself, such as loess hills. Although very thin (e.g., ≤ 25 cm) surficial sediments are prone to mixing by normal pedologic processes, to the point that they lose their identifying depositional morphology, they can still have an important influence on the soil. In this case, these materials may be identified even though their geomorphic process is not recognized overtly. Loess-influenced colluvium is an example.

Some geomorphic descriptive systems attempt to capture both primary and secondary geomorphic processes at a site. This approach, which may be appropriate for mapping geomorphology, can become complex and confusing when addressing soil geomorphology in the context of soil inventory. The main objective of soil inventory is to address the geomorphic processes and products that directly or substantively influence soils and soil behavior. Larger scale or deep-seated geomorphic processes that do not directly or substantively affect soils, such as deep-seated tectonic or structural phenomena, may be beyond the scope of soil inventory. This information can be included with the physiographic information or in general discussions.

Generic vs. Specific Geomorphic Terms

There is considerable range in the specificity of geomorphic terms. Some terms are very generic (e.g., uplands) and can be useful especially at the coarsest scales. While technically correct, generic terms convey relatively little information. Other terms are more specific (e.g., fault-block mountains). As a general rule, it is better to be more explicit than less. A specific term is more informative than an equally correct but more generic term. For example, "loess hill" provides more information than "hill".

Nested Features

The focus in geomorphic description is commonly the land feature (i.e., a single landform or a dominant landscape) that is most critical to soils and that conveys the most relevant context of that site or area. An example is "Alpha soil occurs on a dune." The foremost feature should be the one that most directly impacts or defines the soil. In some environments, there are multiple landforms of different scales that are each relevant to soil behavior and important in documenting an area. For these, multiple landforms can be used in sequence (nested) to convey important setting information. An example is "Alpha soil occurs on a dune on a stream terrace." If nested terms are used, typically no more than two or three are necessary.

Anthropogenic Features

Historically, soil survey in the United States has focused on natural processes, associated sediments, and resulting surface features, such as landscapes, landforms, and microfeatures. Human-altered surface features and materials were traditionally excluded or minimally recognized. They were considered to be artificial phenomena, unpredictable in composition and occurrence, and largely outside the scope of natural process-based soil survey. In recent years, however, there has been a growing awareness and acceptance of the impact of humans upon natural systems and associated features and materials. Anthropogenic features and materials differ from natural phenomena in their origin and processes of formation but can be surveyed in ways similar to those used for conventional geomorphic entities. For example, they can be identified by recurring surface expression (form and arrangement), range of composition and internal arrangement, and lateral extent. As with natural landforms and materials, the ability to consistently partition anthropogenic features into meaningful subsets facilitates recognition of anthropogenic soil geography and greatly assists in land management decisions that concern them.

Terms describing anthropogenic features were adopted by the NCSS in 1993. A new geomorphic category was established to accommodate human-altered or -created features of all scales and to elevate their recognition to the same stature as natural features. Since then, appreciation of the extent of human alterations of the Earth's surface has continued to evolve. As the number and variety of recognized anthropogenic features increased, proposals were made to divide anthropogenic features into three subsets loosely analogous to partitions of naturally derived earth surface features. Furthermore, the International Committee on Anthropogenic Soils defined the phrase "anthropogenic feature" in a non-geomorphic context (any artificial artifact, mark, mold, impression, etc). Subsequently, "anthropogenic features" (geomorphic context) has been replaced with three new, loosely scale-dependent categories: anthroscapes, anthropogenic landforms, and anthropogenic microfeatures.

An anthroscape is important both for its evocative simplicity as a term and its explicit recognition of human-modified lands as legitimate and significant areas. These lands are substantially different from natural systems because they have different sediments, arrangements of sediments, and water dynamics and subsequently require different management practices. An anthroscape is a human-modified "landscape" of substantial and permanent alterations formed by the removal, addition, or reorganization of the physical shape and/or internal stratigraphy of the land. It is associated with management for habitation, commerce, food or fiber production, recreation, and other human activities that have substantively altered water flow and sediment transport across or within the regolith. Types of anthroscapes include urban, suburban, reclaimed land, and agricultural.

An anthropogenic landform is a discrete, human-made "landform" on the Earth's surface or in shallow water that has an internal composition of unconsolidated earthy, organic, human-transported materials, or rock. It typically has straight line boundaries or geometric shape. It is the direct result of human manipulation or activities. It can be mapped at common soil survey scales, such as order 2 (> 1:10,000 to < 1:24,000). Anthropogenic landforms can originate from deposition (e.g., an artificial levee) or removal.

Quarries are an example of an anthropogenic landform.

An anthropogenic microfeature is a discrete, individual, human-derived form on the Earth's surface or in shallow water that has a range in composition of unconsolidated earthy, organic, human-transported materials, or rock. It typically has a recognizable human-imposed shape. It is the direct result of human manipulation or activities. It typically cannot be mapped at common soil survey scales, such as order 1 (< 1:10,000) but can be observed locally. Anthropogenic microfeatures can originate from deposition or removal.

Soil conservation terraces are an example of an anthropogenic microfeature.

Ideas of anthropogenic features will continue to evolve and grow in coming years. The proposal to recognize a new geologic age the Anthropocene continues to gain support and reflects this overall trend. Information on composition, occurrence, and behavior of anthropogenic features and materials, despite their unique differences from "natural" geomorphic phenomena and processes, can assist in wise land management.

Surface Morphometry

Surface morphometry uses various terms to describe land surface shape or geometry, discrete portions of a geomorphic entity or slope segment, and miscellaneous features that are fundamental to soil and natural resource inventory. Several terms are discussed in the following paragraphs.

Elevation is the height of a point on the Earth's surface, relative to mean sea level. This information is widely available from common GIS databases and historically from topographic maps. Elevation conveys the important climatic context and reflects the relative potential and kinetic energy available at a location.

Soil slope has a scale connotation. It refers to the ground surface configuration for scales that exceed about 10 meters and range up to the landscape as a whole. It has gradient, complexity, length, and aspect. The scale of reference commonly exceeds that of the pedon and should be indicated. It may include an entire map unit delineation, a soil component within the map unit delineation, or an arbitrary area. Most commonly, slope is recorded in pedon descriptions for the segment of the landscape extending a few tens of meters above and below the site of the soil profile described and is representative for the landscape segment occupied by the soil component at that site.

Slope aspect is the compass bearing that a slope faces looking down slope. It is recorded either in degrees, accounting for declination, or as a general compass orientation. The direction is expressed as an angle between 0 and 360 degrees (measured clockwise from true north) or as a compass point, such as east or north-northwest.

Aspect can substantially impact local ecosystems. The impact generally increases as slope gradient and latitude increase. In the mid-latitudes of the conterminous United States, this effect becomes particularly important on slopes of approximately 6 to 8 per cent or greater. Increased or decreased solar radiation on slopes due to aspect can affect water dynamics across a site. In the northern hemisphere, north-northeast aspects reduce evapotranspiration and result in greater soil moisture levels, improved plant growth and biomass production, higher carbon levels, and improved drought survival rates for plants. Increased solar radiation on south-southwest aspects increases evapotranspiration and decreases biomass production, seedling survival rates, and drought survival rates for plants.

Effect of slope aspect on vegetation and tree seedling survival.

Slope gradient is the inclination of the land surface with respect to the horizontal plane. It is also commonly referred to as "slope percent" or simply "slope." It is calculated as the vertical distance divided by the horizontal distance ("rise over run"), multiplied by 100, determined at a point along a line oriented up and down slope. It directly controls the kinetic energy, erosive power, and sediment carrying capacity of running water (as overland flow or channel flow), all of which increase with increasing gradient. It inversely affects the amount of time that internal soil water is present. Many soil conservation practices, such as conservation terraces, are designed primarily to reduce slope gradient to minimize soil erosion and increase infiltration. Slope gradient also directly affects land management practices by limiting ranges of operation for various types of equipment, such as tractors and log skidders.

Slope complexity is the relative linearity or smoothness (simple) or irregularity (complex) of the ground surface leading down slope and through the point or map unit of interest. Simple slopes allow the maximum slope length with comparatively unimpeded slope wash processes. In contrast, complex slopes are composed of a series of steps commonly associated with bedrock-controlled benches or other stepped surfaces. These localized breaks in slope reduce slope length, alter slope wash processes, and commonly correspond to changes in soil types.

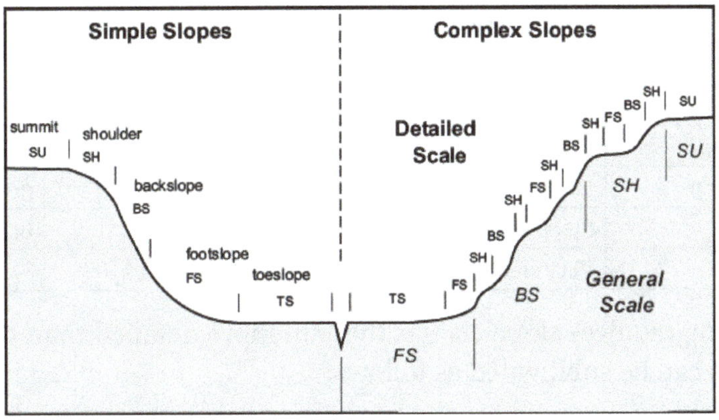

Simple versus complex slopes and slope positions. Note the choice of a detailed or general scale approach to slope position for complex slopes (right side).

In many places, internal soil properties are more closely related to the slope complexity than to the gradient. Slope complexity has an important influence on the amount and rate of runoff and on sedimentation associated with runoff. It can also affect soil temperature through local variation in soil aspect.

Complex slopes on a hillslope of interbedded sedimentary rocks in Wildcat Hills, Nebraska.

Traditionally, slope (gradient) classes are assigned to soil map units to convey the dominant range of slope gradients occurring within it. The numerical slope class limits of map units are not always consistent within or between survey areas. They can vary from one survey area to another, to better capture the local survey slope conditions, as long as they generally remain within the maximum (upper) and minimum (lower) class limits. Descriptive adjectives corresponding to specified slope ranges can be used in text. Such adjectives are slightly different for the mid-range slope classes, depending upon whether the dominant slopes are simple or complex. Gently sloping or undulating soil map units, for example, can be defined with slope class ranges as broad as 1 to 8 percent or as narrow as 3 to 5 percent. Classes may exceed the broadest range indicated in table 2-3 by one or two percentage points where the range is narrow and by as much as 5 percent or more where the range is broad. The slope class terms can also be used in naming slope phases of map units.

Table: Definitions of Slope Classes.

Classes for		Recommended slope (gradient) class limits	
Simple slopes	Complex slopes	Lower (percent)	Upper (percent)
Nearly level	Nearly level	0	3
Gently sloping	Undulating	1	8
Strongly sloping	Rolling	4	16
Moderately steep	Hilly	10	30
Steep	Steep	20	60
Very steep	Very steep	> 45	

If the detail of mapping requires slope classes that are more detailed than those in table some or all of the slope classes can be subdivided as follows:

- Nearly level: Level, nearly level.
- Gently sloping: Very gently sloping, gently sloping.
- Strongly sloping: Sloping, strongly sloping, moderately sloping.
- Undulating: Gently undulating, undulating.
- Rolling: Rolling, strongly rolling.

In a highly detailed survey, for example, slope classes of 0 to 1 percent and 1 to 3 percent would be named "level" and "nearly level," respectively.

Slope length is rarely used directly in soil mapping because its range across all the polygons of a soil map unit is highly variable. Furthermore, natural slope lengths are commonly interrupted and artificially shortened by human-made features such as ditches, roads, or field boundaries. Slope length does have important uses in key soil erosion programs and models, including the Revised Universal Soil Loss Equation, version 2. It has considerable control over surface water runoff and potential for accelerated water erosion. Generic terms such as "long" or "short" can be used to describe slope lengths that are typical of certain kinds of soils. These terms are typically relative within a physiographic region. A "long" slope in one place might be considered "short" in another. If such terms are used, they are defined locally. For observations at a particular point, it may be useful to record the length of the slope that contributes water to that point (called point runoff slope length) as well as the total length of the slope. The sediment transport slope length is the distance from the expected or observed initiation point upslope of runoff to the highest local elevation where deposition of sediment is expected to occur. This distance is not necessarily the same as the point runoff slope length.

Relative slope segment position indicates vertical subdivisions of long slopes. It can be useful, especially in areas of substantial slope length, to identify general "slope segment" positions, such as lower third, middle third, and upper third. For example, the long slopes of mountainflanks commonly exhibit changes in bedrock stratigraphy somewhere along the slope that correspond to soil types that differ in parent material composition and type and amount of rock fragments.

Slope shape is the dominant form of the ground surface curvature. It is expressed in two directions, which are paired up and down slope (vertical, or perpendicular or normal to the slope contour) and across slope (horizontal, or along the slope contour). When used in tandem, the slope directions describe the configuration of the surface of a portion of the slope and the soil upon it. Both slope directions can be described by one of three curvature shapes: convex, linear, or concave. In the up and down direction, the surface of a linear slope is substantially a straight line when seen in profile at right angles to the contours. The gradient neither increases nor decreases significantly with distance. An example is the dip slope of a cuesta. On a concave slope, gradient decreases down the slope. An example is a foot slope. Where the slope decreases, runoff water decelerates and tends to deposit sediments, as on the lower parts of the hill slope. Simultaneously, as surface water flow slows, it has greater opportunity to infiltrate into the soil. On a convex slope, such as the shoulder of a hill or ridge, gradient increases down the slope and runoff tends to accelerate as it flows down. If contours are substantially straight lines (parallel), as on the side of a lateral moraine, the across slope shape is linear. An alluvial fan has a convex contour that bows outward. A cirque has concave contours. In figure, nine possible combinations of linear (L), concave (C), and convex (V) slopes are shown. For both the up and down orientation and the across slope orientation, where the slope is convex, surface runoff water tends to diverge (spread apart) as it moves down the slope. As a result, overland flow is dissipated and both the erosive power and the amount of water available for infiltration are reduced. Where the slope is concave, surface runoff water tends to converge, or concentrate. The most intense concentration of running water occurs where both orientations are concave, as in a swale on a hillside or in a head slope at the head of a drainageway. The most intense divergence of running water occurs where both orientations are convex, as on a nose slope at the end of a ridge.

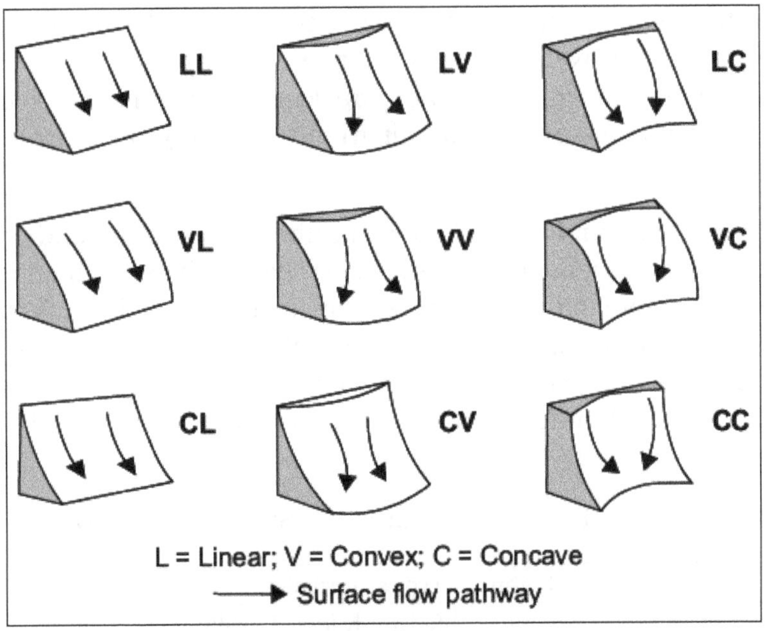

Slope shape based on combinations (up and down slope and across slope) of surface curvature.

Table: Geomorphic Component Terms for Hills.

Geomorphic component term	Typical attributes
Interfluve	High, relatively level area that generally does not receive run-on surface flow; residuum, short-transport colluvium.
Crest	High, narrow area; converging backwearing slopes that form a lowered ridge.
Head slope	Convergent overland water flow; thickened colluvium, slope alluvium.
Side slope	Parallel overland waterflow; colluvium, slope alluvium, pedisediment, residuum.
Nose slope	Divergent overland water flow; colluvium, slope alluvium, pedisediment.
Free face	Rock outcrop.
Base slope	Concave surface; colluvium, slope alluvium.

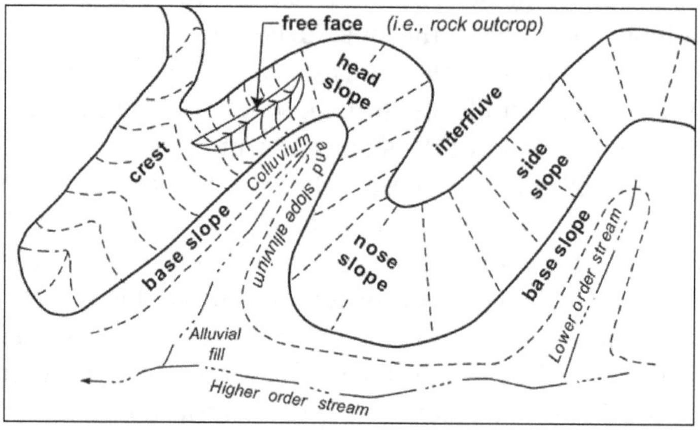

Three-dimensional depiction of geomorphic components of hills.

Hillslope profile positions (also called hillslope positions) refer to two-dimensional segments of a line used to describe slope position along a transect oriented up and down slope (normal to the slope contour). They do not address lateral dimensions. These line segments, progressing from the top of the slope to the bottom, are: summit, shoulder, backslope, footslope, and toeslope. These terms have proven useful for many decades because they can describe areas on slopes where soil bodies are consistent and breaks in slope curvature where soils typically change. They can be used alone or in a combination to verbally capture where soils recur up and down slopes.

Geomorphic components are similar to hillslope profile positions (up and down slope) but include an additional lateral dimension (across slope) that enables distinctions to be made between the slope curvatures of land areas in three dimensions. They indicate patterns of surface water flow, such as concentration, dispersion, or parallel (lateral) flow. Not all settings, however, share the same, recurrent configurations. For this reason, geomorphic component descriptors have been developed for four different settings: hills, terraces and stepped landforms, mountains, and flat.

Table: Geomorphic Component Terms for Terraces and Stepped Landforms.

Geomorphic component term	Typical attributes
Tread	Relatively level, broad surface.
Riser	Vertical or steep side slope separating treads.

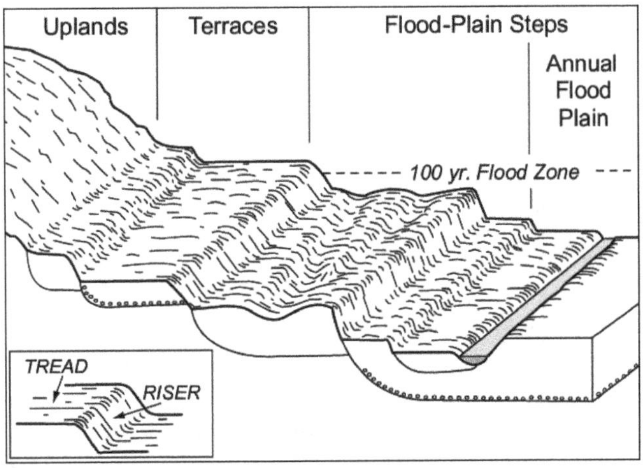

Three-dimensional depiction of geomorphic components of terraces and stepped landforms.

Table: Geomorphic Component Terms for Mountains.

Geomorphic component term	Typical attributes
Mountaintop	High area (crest, summit); residuum or short transport colluvium, solifluction deposits; generally does not receive run-on surface flow.
Mountainflank	Complex slopes, long slopes, substantial gradients, colluvium, mass wasting deposits, talus.
Free face	Rock outcrop.
Mountainbase	Concave surface; thick colluvium, slope alluvium, mass wasting deposits.

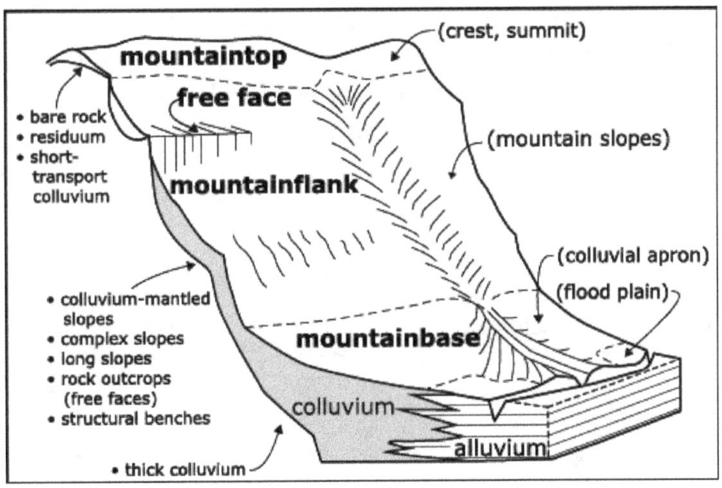

Three-dimensional depiction of geomorphic components of mountains.

Table: Geomorphic Component Terms for Flat Plains.

Geomorphic component term	Typical attributes
Rise	Slightly elevated area (1-3% slopes)
Talf	Very low slope gradients (0-1%); deranged or incipient drainage network; lacustrine deposits, alluvium, till, marine deposits, eolian deposits, and other flat-lying deposits
Dip	Depressions; backswamp deposits, marl, organic deposits, and other deposits in low-lying areas

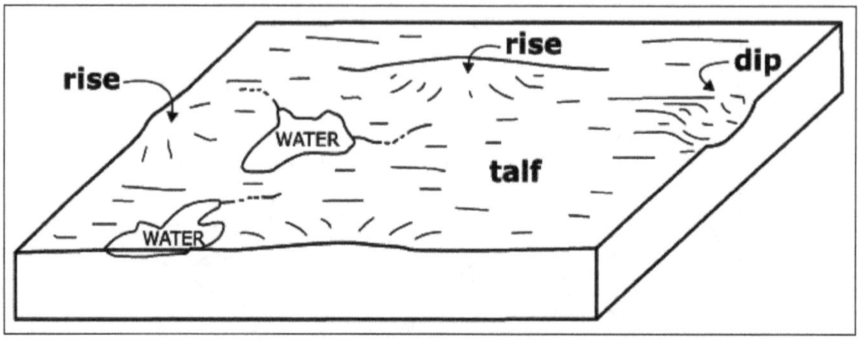

Three-dimensional depiction of geomorphic components of flat plains.

As with hillslope profile positions, geomorphic components for hills have been widely used in one form or another. These terms and concepts work very well for partitioning and describing hilly terrain as functionally distinct members. However, these same concepts work poorly when applied to very gentle terrain. The kinetic energy of running water is dramatically and functionally less at low gradients. The erosive power of the water is reduced as well as its sediment carrying capacity, which determines what sediments are removed and which are left behind. Additionally, there is a general increase, compared to higher gradient systems, in the residence time of water, particularly internal soil water, which alters the biogeochemical dynamics and products. Therefore, new concepts and associated terms were developed for geomorphic components of flat plains. In a similar way, hillslope components were found to be inadequate when applied to high-gradient terrain. Very long and commonly complex slopes and much greater kinetic energy dramatically increase erosion potential and the sediment carrying capacity, can change sediment winnowing effects, can decrease soil water residence time, and can otherwise alter system dynamics and resulting sediments and soils. Therefore, new concepts and associated terms were developed for geomorphic components of mountains. Stream terrace and stepped landforms are also sufficiently unique to warrant separate geomorphic component descriptors.

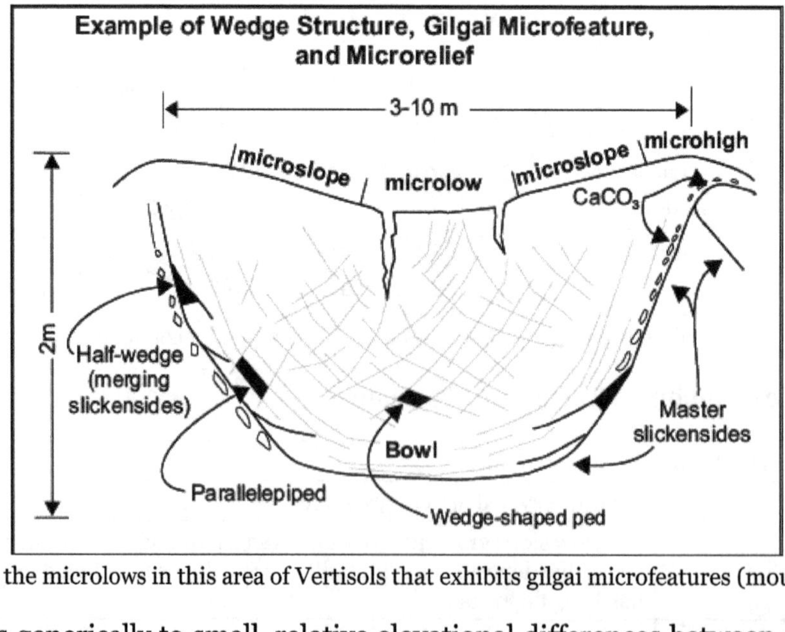

Water ponds in the microlows in this area of Vertisols that exhibits gilgai microfeatures (mounds and bowls).

Microrelief refers generically to small, relative elevational differences between adjacent areas on the earth surface. In subaerial settings, minor elevational differences can profoundly influence plant growth above ground and, subsequently, water conditions below ground. The lateral scale across which the elevational differences occur is generally on the order of about 3 to 10 meters but can be smaller. A gilgai, which has micro site differences in patterned ground, is an example. Terms used to describe microrelief positions are microhigh, microslope, and microlow.

Drainage patterns (also called a drainage network) describe the recurring arrangement of interconnected drainage channels across the surface of a land area. They provide substantial insight into the underground, controlling bedrock or regolith as well as the locations where these materials and the overlying soils change. Figure presents the more general patterns. Drainage patterns can be best observed and evaluated on aerial photographs, topographic quadrangle maps, or GIS spatial layers that present drainage way patterns in detail.

Table: Illustrations and descriptive terms for drainage patterns.

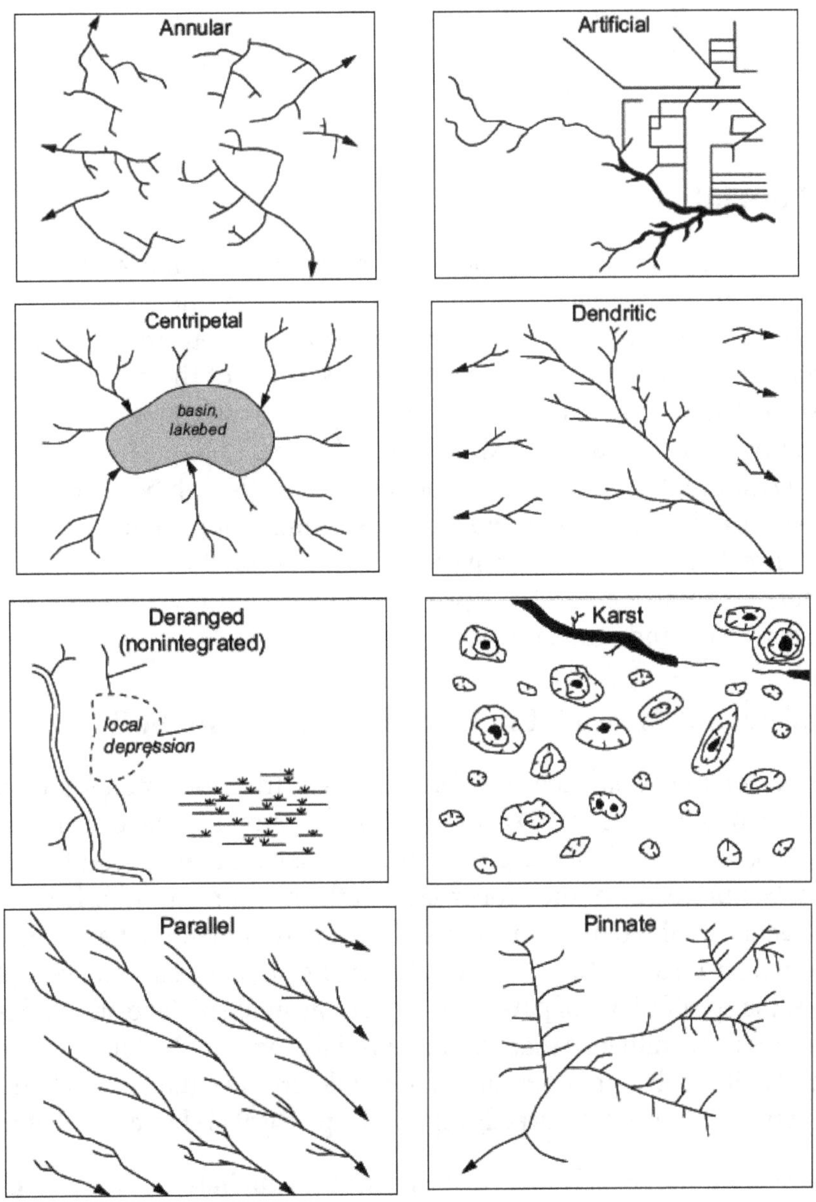

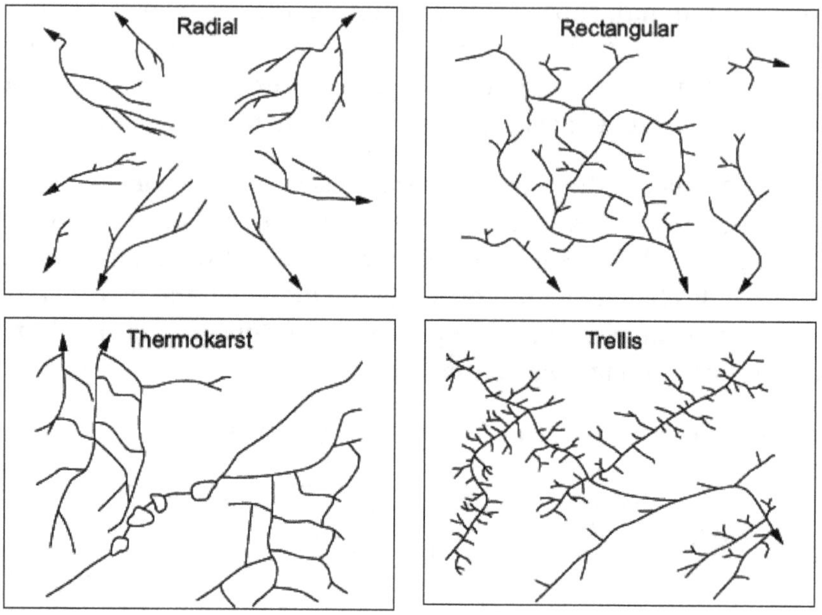

Soil Parent Material

Soil parent material refers to the unconsolidated, organic and mineral materials in which soils form. The unconsolidated material, or regolith, in which a soil develops exerts tremendous influence upon what that soil is and is not and how it behaves. Determining the parent material is therefore important in accurately identifying the composition of the soil. Parent material is more than just soil texture. Other attributes, such as mineralogy, stratigraphy, and the degree of sorting and particle rounding, can substantially affect soil behavior. Eolian sand, such as dunes, can behave hydrologically different than beach sand deposits, even though both are made of sand, due to differences in the internal arrangement and lateral continuity of primary particles. Accurate identification conveys direct and implicit information about the soil itself, the environment in which it formed, and its current environment. Soils provide a record of prevailing and past environments, climates, human activities, and much more.

Importance of Parent Material in Understanding the Soil

Soil formation involves alterations, such as additions, losses, transformations, and translocations and including weathering, of unconsolidated earthy or organic materials. The parent material of a genetic soil horizon cannot be observed in its original state as it has undergone soil formation. Rather, the original state must be inferred from the properties that the horizon has inherited and from other evidence, such as the geomorphic context. In some soils, the parent material has changed little and what it was originally can be deduced with confidence. In other soils, such as some very old, highly altered soils of the Tropics, the specific kind of parent material or its mode of deposition is less clear and more speculative. Regardless, the influence that parent material exerts on the inherent properties and subsequent behavior of soil is substantial. Parent material determines the broad characteristics of what is geochemically present or absent. It directly affects the physical architecture that makes up a soil.

Much of the mineral matter in which soils form is derived from hard bedrock in some way. Glaciers may grind the bedrock into fragments and smaller particles and deposit the unsorted mixture as till. Wind and running water can abrade and entrain small particles that accumulate elsewhere as eolian or fluvial deposits. Bedrock may be weathered and significantly changed chemically and physically but not be moved from its place of origin. Little may be gained from attempting to differentiate between geologic weathering and soil formation because both are weathering processes. It may be possible to infer that a material was weathered before soil formation. The weathering process causes some bedrock constituents to be lost, some to be transformed, and others to be concentrated.

Soil parent material is not always residuum weathered directly from underlying bedrock. The material that developed into the modern soil may not be related to the underlying bedrock at all. In fact, most soils did not form in place but were subject to transport and deposition by wind, water, gravity, or human activities.

Seldom is there absolute certainty that a highly weathered material actually weathered in place. The term "residuum" is used if the properties of the soil indicate that it has been derived from rock similar to that which underlies it and if there is no overt evidence that it has been modified by movement. A decrease in the amount of rock fragments as depth increases, especially over saprolite, indicates that soil material probably has been transported down slope. Stone lines, especially if the stones have a different lithology than the underlying bedrock, are evidence that the soil did not form entirely in residuum. In some soils, transported material overlies residuum and illuvial organic matter and clay are superimposed across the discontinuity between the contrasting materials. A certain degree of landscape stability is inferred for soils that formed in residuum. A lesser degree is inferred for soils that developed in transported material.

Standard terms are used to describe both consolidated and unconsolidated materials beneath the solum that influence the genesis and behavior of the soil. Besides primary observations, the scientist uses his own judgement to infer the origin of the parent material from which the solum developed. Primary observations must precede, and be clearly separated from, inferences.

The lithologic composition, structure, and consistence of the material directly beneath the solum are important. Evidence of stratification of the material should be noted. It includes textural differences, stone lines, and changes in kind and amount of coarse fragments. Commonly, the upper layers of outwash deposits settled out of more slowly moving water and are finer in texture than the lower layers. Windblown material and volcanic ash are laid down at different rates in blankets of varying thickness. Examples of such complexities are nearly endless.

Where alluvium, eolian sands, volcanic ash, or colluvium is rapidly deposited on old soils, buried soils may be well preserved. In other places, the accumulation is so slow that the thicknes of the solum increases only gradually. In these places, the material beneath the solum that was once near the surface may now be buried below the zone of active change.

Where hard rocks or other strongly contrasting materials lie close enough to the surface to affect soil behavior, their properties and the depth to contact need to be measured accurately. The depth of soil over such nonconforming materials is an important criterion for distinguishing different kinds of soil.

General Kinds of Parent Material

Broad groupings of parent material are discussed in the following paragraphs. Consistent use of terminology to describe parent materials in pedon descriptions and databases enhances the usefulness of the information and allows easier and more reliable comparison of soils that formed in the same kind of parent material. The NCSS has adopted standard terms for many kinds of parent material.

Material Produced by Weathering of Bedrock

The nature of the original rock affects the residual material produced by weathering. Bedrock undergoes various changes as it weathers, beginning with the progressive removal of readily weatherable minerals, such as plagioclase feldspar and biotite mica. The relative ease of weathering of major minerals was described by Goldich and refined for some soil clay minerals by McClelland. This weathering sequence indicates which minerals weather most readily and the relative order in which weathering progresses. Evaluating which minerals are present and which have been removed can indicate the degree of weathering that rock has undergone.

In-place deposits. Saprolite is soft, friable material produced by bedrock that has been highly weathered in place (in situ). The weathering process has removed mineral constituents but left the fabric and structure of the original rock without significant loss in volume. If the altered material has lost most or all rock fabric and structure and its original volume has been reduced (e.g., by void collapse), the unconsolidated, in-place earthy material is called residuum. Such distinctions are useful in recognizing close geochemical and physical relationships to the bedrock of origin. It is assumed that residuum is in situ and has not undergone substantive lateral displacement or transport. Residuum is a major kind of parent material, particularly on older, stable landscapes and in warm and humid climates.

If the soil is derived directly from underlying bedrock and exhibits little or no evidence of lateral transport, the parent material should be identified (e.g., residuum, grus, saprolite, bauxite) and then paired with the kind of bedrock from which it was derived. The point where rock weathering ends and soil formation begins is not always clear. The processes may be consecutive or overlapping. Quite different soils may form from similar or identical rocks under different weathering conditions. Texture, color, consistence, and other characteristics of the parent material should be included in the description of soils, as well as important remnant bedrock features, such as quartz dikes. Information about the mineralogical composition, consistence, and structure of the parent rock is useful and should also be included.

Transported Material

Most soil parent materials have been moved from their place of origin and deposited elsewhere. The principal subsets of transported materials are typically arranged according to the main geomorphic process responsible for their transport and deposition. In most places, there is sufficient evidence to make a clear determination.

In soil morphology and classification, it is very important to observe and describe the characteristics of the parent material. It is not enough simply to identify the material. Any doubt regarding the identification should be mentioned. For example, it can be difficult to determine if silty deposits are alluvium or loess or to distinguish silty colluvium from silty residuum. It can also be difficult to distinguish certain mud flow deposits from till or to distinguish some sandy tills from sandy outwash. Additional observations across large exposures or at multiple locations help in making such distinctions. These distinctions provide supporting information needed to accurately inventory soils and thereby improve the accurate prediction of soil behavior.

Water-laid or water-transported deposits. Alluvium is a widely occurring parent material. It consists of unconsolidated, sorted, clastic sediment deposited by running water, particularly channel flow. It may occur on actively flooded portions of modern streams. Remnants of old stream terraces may occur in dissected areas far away from or high above, a present stream or occur as paleoterraces that are unrelated to the modern stream. In larger streams and rivers, a series of alluvial deposits in the form of stream terraces may loosely parallel the modern stream. The youngest deposits occur in the stream; deposits increase in age as they progress to higher levels. In some areas, recent alluvium covers older terraces. For example, younger alluvial fan sediments onlap and bury older fan sediments. Alluvium is also the dominant parent material in large tectonic valleys, such as the bolsons and semi-bolsons of the Basin and Range Physiographic Province in the western United States. On these broad, sloping landscapes, alluvium occurs as thick deposits on active alluvial fans and fan remnants or as broad, relatively level alluvial flats on basin floors. The further down a river system alluvium occurs, the better sorted the sediments tend to be. Larger stream systems commonly have backswamp deposits along low-gradient stream reaches. These lower energy areas are set back from the main channel and are dominated by sediments that are laminated and finer (silts and clays) than alluvium closer to the stream channel. Slope alluvium refers to hillslope sediments transported primarily by slope wash processes (sheet flow) rather than by the channel flow of streams. Crude lateral particle sorting is evident on long slopes, but it is much less evident than the particle sorting in alluvium derived from channel flow.

Lacustrine deposits consist of clastic sediments and chemical precipitates that settled out of bodies of still water, such as ponds and lakes. Deposits associated directly with glaciers and laid down in freshwater lakes (glaciolacustrine deposits) or in oceans (glaciomarine deposits) are included with other glacial deposits. Numerous basins in the western U.S. contained moderate to large pluvial lakes during the Pleistocene epoch. These lakes have either drastically shrunk or disappeared during the warmer and drier climates of the Holocene epoch. The now dry lakebeds are known as playas or salt flats and contain thick lacustrine deposits dominated by silt and clay with inter bedded layers of volcanic ash. Some also contain substantial evaporate deposits. Soils in the narrow margins of these barren playas are generally saline, depending on climate and drainage, and are sparsely vegetated with salt-tolerant plants.

Marine deposits settled out of the sea, lagoons, or estuaries and commonly were reworked by currents and tides. Subaqueous soils include sediments that remain under water. Some marine deposits were later exposed either naturally by falling sea levels or following the construction of dikes and drainage canals. Many of the soils of the Atlantic and Gulf of Mexico Coastal Plains in the south-eastern U.S. formed in marine sediments deposited during a time of higher sea level. These deposits vary widely in composition. In low-energy settings, such as lagoons, sediments tend to be finer textured and may have intermittent or substantial amounts of organic materials. Higher energy settings can have substantial amounts of sandy material (such as in areas of inlets and barrier islands) or coarser rock fragments (such as in areas of rocky coasts and headlands).

Beach deposits mark the present or former shorelines of lakes or oceans. They consist of low sheets or ridges of sorted material. They are commonly sandy or gravelly (along non-rocky coasts) or cobbly or stony (especially along rocky coasts).

Eolian deposits are very well sorted windblown material. They are broadly divided into groups based on dominant particle size or origin. Examples are aerosols, dust, loess, and eolian sands. All but the finest wind-driven sediments share some depositional traits. "Lateral fining" refers to the progressive reduction in average particle size and deposit thickness as distance increases along the prevailing wind direction and away from the source area. The closer to an eolian sediment source (e.g., a large barren flood plain), the coarser the average particle size and the thicker the eolian deposit. The dominant particle sizes of discrete eolian deposits range from silt and very fine sands (loess) and from fine to medium sands (eolian sands).

Eolian sands are significant due to their physical prominence and the wide range of distinct landforms (especially dune types) they produce. Very fine and fine eolian sands commonly occur as dunes and medium sands tend to form sand sheets. Eolian sands are common in, but not limited to, warm, dry regions. They characteristically consist of sands with a high content of quartz and a low content of clay-forming materials. Sand dunes may contain large amounts of calcium carbonate or gypsum, especially in deserts and semi-deserts.

During periods of drought and in deserts, local wind movements may mix and pile up soil materials of different grain sizes, including materials with a high content of clay. Sand-sized aggregates of clay (e.g., parna) can even form dunes (parna dunes). In areas where sand and finer eolian materials are intimately intermingled, the eolian materials may be identified generically as eolian deposits rather than as distinct loess or eolian sands.

Loess deposits are important because their physical and mineralogical properties make them highly suitable to food and fiber production worldwide. Their texture is typically very silty but may range from fine silt to very fine sand. Most loess is pale brown to brown, although gray and red colors are also common. Some colors are inherited from the source material (geogenic colors). Other colors, particularly gray colors, may be caused by post-deposition soil formation, such as redoximorphic alteration resulting in iron reduction. Although thick loess deposits appear to be relatively massive, they have some gross vertical cracking with coarse polygonal structure and can support nearly vertical walls (e.g., roadcut walls) for many years. Silty deposits that formed in other ways have some or all of these characteristics. Windblown silt that has been leached and strongly weathered can be acidic and rich in clay, whereas some young deposits of loess that are mainly silt and very fine sand have a low content of clay.

Other, finer windborne particles also affect soils in unique ways but are not generally recognized as a kind of parent material. Dust is composed of clay or very fine silt-sized particles and can be deposited dry or in precipitation. It can travel great distances from its point of origin, even circle the Earth in the upper atmosphere and be deposited in small increments across the world. After dust settles, the very fine particles are readily mixed into pre-existing soils and may substantially affect soil properties. However, they typically do not form readily identifiable, discrete deposits by themselves. Aerosols are the finest of particulate materials, so small that they can stay suspended in air for extended periods. Wood ash is an example. These particles are typically too fine and too diffuse to accumulate as separate deposits. Consequently, they are not identified as discrete parent materials in soil survey. Nonetheless, they can contribute meaningful amounts of carbon ash, pollen, quartz, or other materials to soils. They typically settle out as raindrop nuclei and infiltrate soil in suspension or settle in water bodies. Other soil constituents accompany precipitation, such as atmospheric elements in solution (fixed nitrogen, sulfur, calcium, magnesium, sodium, potassium, etc.) but are not included within the concept of parent materials.

A conventional practice in considering geomorphic processes is to include volcanic eolian deposits, such as ash and pumice, with other volcanic materials because their origin, including mineralogical composition and depositional dynamics, is closely associated with volcanism.

Glacial and periglacial deposits are derived from material moved and deposited by glacial processes or associated with cold climates. However, the two types have two distinct geomorphic process systems. Their processes and sediments are commonly associated because they share very cold climatic settings and driving forces. They are considered together here for convenience. Glacial refers to materials that have been directly created, moved, and deposited by glacial ice (i.e., drift and till). A conventional practice in considering geomorphic processes is to include glaciofluvial, outwash, and glaciolacustrine deposits among other glacial materials because their origins, including depositional dynamics, resulting stratigraphy, and mineralogical composition, are closely associated.

Drift is a general, inclusive term for all material picked up, mixed, disintegrated, transported, and deposited by glacial ice or glacial melt waters. The term is so generic that it is principally used for very coarse scales that prohibit details. In many places, drift is mantled by loess. Thick mantles of loess are typically easily recognized, but very thin mantles may be so mixed by soil-building processes that they can scarcely be differentiated from the underlying drift.

Till is a type of drift that was deposited directly by ice and had little or no transportation by water. It is generally an un-stratified and heterogeneous (i.e., unsorted) mixture of clay, silt, sand, gravel, and boulders. Some of the ice-entrained mixture settled out as the ice melted and was subject to very little washing or reworking by water (ablation till), and some was overridden by the glacier and became compacted (lodgement till). Till occurs in various glacial landforms. Ground moraines and recessional moraines are examples. In many places, it is important to differentiate tills of several glaciations. Commonly, the tills underlie one another and may be separated by other deposits or old, weathered surfaces. In many cases, till was later eroded by the wave action in glacial lakes. The upper part of such wave-cut till may have a high percentage of rock fragments.

Till ranges widely in texture chemical composition and degree of weathering. It is principally affected by the composition of the bedrock it has overridden and whose materials it has entrained. Tills of the mid-continental U.S. are underlain by sedimentary rocks, such as limestone and shale, and typified by heavy textures (clay, clay loams). In contrast, tills of northern Minnesota, New England, and Canada underlain by crystalline bedrock, such as granite, are typified by coarser textures (gravelly sandy loam). Much till is calcareous, but a significant amount is non-calcareous because no carbonate rocks contributed to the till or because subsequent leaching and chemical weathering have removed the carbonates. The two most widely occurring and operationally important types of till are ablation till and lodgement till. Ablation till is characterized by a comparatively low bulk density (e.g., 1.4 g/cm^3) and occurs at the top of till deposits. Lodgement till formed beneath a glacier and was over-compacted. As a result, it has a very high bulk density (e.g., 1.8 g/cm^3) that substantially restricts internal water flow and makes excavation difficult. Some tills are identified by position of formation relative to the glacial ice. Supraglacial till formed by the sediments on top of or entrained with the ice that settled out as the ice melted (ablation till or melt-out till) or moved as localized mud flows (flow till). Subglacial till, such as lodgement till, formed beneath glacial ice.

Glaciofluvial deposits are materials moved by glaciers and subsequently carried, sorted, and deposited by melt waters flowing from the ice. Outwash is a parent material term for the detritus (chiefly sand and gravel) removed or "washed out" from a glacier by melt water streams and deposited beyond the ice front or end moraine. The coarsest material was deposited nearer the ice. This outwash commonly forms on plains, valley trains, outwash terraces, or deltas in drainage ways or in relict glacial lakes. Some outwash terraces may extend far beyond the farthest advance of the ice. Near moraines or in disintegration moraine landscapes, sorted glaciofluvial material may form kames, eskers, and crevasse fills.

Glacial beach deposits consist of rock fragments and sand. They mark the locations of relict shorelines (i.e., strandlines) of former glacial lakes. Depending on the character of the original drift, beach deposits may be sandy, gravelly, cobbly, or stony.

Glaciolacustrine deposits are also derived from glaciers but were reworked and laid down in glacial lakes. These deposits range from fine clay to sand. Many of them are stratified or varved. A varve is the pair of deposition laminae for a calendar year. The finer portion reflects lower energy deposition during the cold season, and the slightly coarser portion reflects higher energy deposition during the warmer season when runoff is greater and wave action occurs.

In many places, it is difficult to distinguish between the different kinds of glacial sediments. For example, pitted outwash plains can be difficult to distinguish from sandy till in recessional moraines and wave-cut till can be difficult to distinguish from lacustrine material. Typically, even the most subtle differences can be identified from multiple, well planned field observations. This information is used to accurately determine the geomorphic setting and its associated sediments. Careful observations and descriptions of parent material, stratification, coarse fragment distribution and the surface forms in which they occur provide hard evidence needed for correct conclusions. However, some situations are not fully understood at present because of their complexity or incomplete scientific knowledge.

Periglacial deposits have several major types. Cryoturbates are deposits of sediments that have been mixed or preferentially sorted by seasonal frost heave, partial melting and refreezing of permafrost, or other non-glacial ice displacement processes. These processes can organize sediments in several ways. Internally, the materials typically exhibit convolutions or low-grade internal sorting, unlike the more horizontal layering typical of mineral soils in warmer climates. Surficial sorting, particularly of coarse fragments, can take the form of polygons or stripes or other patterned ground. Solifluction deposits consist of heterogeneous mixtures of textures, including rock fragments. The orientation of the rock fragments indicates the slow downslope movement that resulted in surficial lobes, sheets, and terraces. Solifluction deposits form in response to seasonal or partial thawing of the near surface "active zone." Periglacial parent materials can have wide aerial extent. Active or recent periglacial deposits occur most extensively at high latitudes or at high elevations outside of, or otherwise unaffected by, glacial ice. Relict solifluction deposits also widely occur in the form of relict patterned ground in association with former continental glaciated areas in mid-latitudes.

Mass wasting (mass movement) deposits are some materials are transported primarily or completely by gravity. Transport can occur extremely quickly or gradually. Landslide deposits are a generic term that includes all forms of landslide materials. These deposits can be more explicitly identified based on the main mode of movement.

Table: Types of Landslide Deposits.

Movement types	Deposit attributes
Fall deposits	Free fall, bouncing or rolling.
Topple deposits	Forward rotation over a basal pivot point.
Slide deposits: Rotational landslide	Backward rotation around a pivot point above the ground surface.
Slide deposits: Translational slide	Mass lateral displacement along a planar slip face.
Spread deposits	Layers plastically extruded by liquefaction.
Flow deposits	Wet or dry mass flow that behaves as a viscous liquid.

Each of these movement types can be further subdivided to indicate the dominant kind of material moved: rock (consolidated bedrock masses), debris (unconsolidated material rich in rock fragments), or earth (dominantly fine-earth material). These terms are useful in specifying different levels of detail needed to identify areas according to their associated deposits. They are also used to convey the composition of the present materials, which impacts land management decisions.

Other kinds of gravity-related deposits are widely recognized. Colluvium is poorly sorted slope sediments that have been transported and accumulated along or at the base of slopes, in depressions, or along small streams primarily due to gravity, soil creep, and slope wash processes. Accumulations of rock fragments at the base of rock outcrops are called talus. Rock fragments in colluvium are typically very angular to sub rounded due to relatively short transport distances and the limited abrasion associated with the process. In contrast, rock fragments in alluvium and glacial outwash are rounded to well-rounded and water worn.

Organic deposits are material dominated by carbon-rich plant or organism detritus. The organic material accumulates more rapidly than it decomposes. This unconsolidated material is commonly associated with, but not restricted to, wet soil or subaqueous conditions. Organic deposits can persist in extremely dry settings or under other conditions that reduce or eliminate microbial decomposition, such as low oxygen or low pH (acidic). These latter conditions can produce various types of organic accumulations that may become the soil parent material generically called "organic materials." Organic deposits can be further defined according to the dominant plant material present, such as woody, herbaceous, grassy, or mossy. Different terms are used to modify an associated soil texture (e.g., mucky, peaty). Terms used in lieu of texture for organic materials include muck, peat, and highly decomposed organic materials.

Some organic materials occur as alternating layers of different kinds that reflect the dominant vegetative cover at the time of deposition. Others are combinations of peat and mineral materials. In some places, organic materials cap, are intimately mixed with, or are discretely interlayered with volcanic ash, marl, alluvium, or eolian sands. Descriptions of organic material should include labels (e.g., woody organic materials) or notations identifying the origin and dominant botanical composition, to the extent that they can be reasonably inferred.

Volcanic eolian deposits, such as ash and pumice, are treated separately from other eolian parent materials because of their unique mineralogy and depositional dynamics. Tephra, volcanic ash, pumice, and cinders are unconsolidated igneous sediments that were ejected during volcanic eruptions and moved from their place of origin. Most have been reworked by wind and, in some places, by water. Tephra is a broad, generic term referring to any form of volcanic ejecta. Various subdivisions are recognized and should be used when possible. Ash is volcanic ejecta smaller than 2 mm. It can be subdivided into fine ash (< 0.06 mm) and coarse ash (> 0.06 and < 2 mm). Pumice is volcanic ejecta larger than ash (> 2 mm) that has a low specific gravity (< 1.0). Cinders are volcanic ejecta larger (> 2 mm and < 64 mm) than ash and heavier (specific gravity > 1.0 and < 2.0) than pumice.

Anthropogenic deposits are human-transported material is a general term for solid phase organic or mineral material that can function as soil or soil-like material. It has been mixed and moved from a source area to a new location by purposeful human activity, usually with the aid of machinery or hand tools. There has been little or no subsequent reworking by wind, gravity, water, or ice. Human-transported materials are most commonly associated with building sites, mining or dredging operations, landfills, or other activities that result in the formation of a constructional anthropogenic landform. Anthropogenic material differs from natural deposits in that its internal composition and stratigraphic arrangements depend upon the emplacement methods, tools, and intentions of people. It is generally more variable and less predictable in its content and configuration than material emplaced by natural processes. Nonetheless, it can be described and broadly quantified in ways similar to how natural materials are evaluated.

In database management, it is helpful of have an alphabetical master list of the many kinds of parent materials. The diverse kinds of parent materials can also be constructively arrayed within subsets based upon the dominant geomorphic processes that erode, transport, or deposit them. Table lists parent material groups based on geomorphic process or setting.

Table: General Groups of Parent Materials Based on Geomorphic Process or Setting.

General groups	Specific examples
Anthropogenic deposits	Dredge spoil, mine spoil, earthy fill
Eolian deposits (nonvolcanic)	Eolian sands, loess
Glacial and periglacial deposits	Till, solifluction deposit
In-place deposits (nontransported)	Residuum, saprolite
Mass wasting deposits	Mudflow deposit, talus
Miscellaneous deposits	Diamicton, gypsite
Organic deposits	Diatomaceous earth, grassy organic materials
Volcanic deposits	Andesitic ash, pumice
Water-laid or water-transported deposits	Alluvium, lacustrine deposit

These subsets compliment and loosely parallel the geomorphic environment categories presented in the Geomorphic Description System used by the NCSS. Soil parent materials should generally relect the dominant geomorphic environment and vice versa.

Multiple Parent Materials

Soil is commonly composed of layers of several different types of parent materials (e.g., colluvium over residuum) that are identifiable in the soil's stratigraphy. For example, till is covered by a mantle of loess in many places. Thick mantles of loess are easily recognized, but very thin (e.g.,< 25 cm) mantles may be so altered by soil-building processes, such as pedoturbation, that they can scarcely be differentiated from the underlying till. The contact between substantially different (contrasting) parent materials in a soil is called a lithologic discontinuity. It should be documented using horizon description nomenclature and other descriptive conventions.

Unconsolidated contrasting soil material may differ in pore-size distribution, particle-size distribution, mineralogy, bulk density, or other properties. Some of the differences may not be readily observable in the field. Some deposits are clearly stratified, such as some lake sediments and glacial outwash, and the discontinuities are sharply defined.

The primary deposition differences of multiple, contrasting parent materials can be confused with the effects of soil formation. Silt content may decrease regularly with increasing depth in soils presumed to have formed in till. The higher silt content in the upper part of these soils can be explained by factors other than soil formation. In some of these soils, small amounts of eolian material may have been deposited on the surface over the centuries and mixed with the underlying till by insects and rodents or freeze-thaw action. In others, the silt distribution may reflect water sorting.

Inferences about contrasting properties inherited from differing layers of geologic material may be noted when the soil is described. Generally, each identifiable layer that differs clearly in properties from adjacent layers is recognized as a horizon or subhorizon. Whether it is recognized as a discontinuity or not depends upon its degree of contrast with overlying and underlying layers and its thickness.

A pragmatic balance is needed between identifying the dominant parent material layer(s) in a soil and not becoming overwhelmed by excessive detail. While there are no rigid criteria, such as a thickness minimum, it is particularly important to identify layers that are physically contrasting enough and thick enough to substantively affect internal water flow. There are several widely recognized exceptions for which numerous sediment layers are not comprehensively described. For deposits that are intrinsically highly stratified, whose lateral continuity is intermittent, it is impractical to identify or sample every thin layer (lamina). For finely laminated alluvium or tephra deposits, only the larger, aggregate layers are identified and sampled as composites (bulked). Minor layers (laminae) within larger layers are noted but typically are not comprehensively documented nor sampled individually.

Bedrock

The term "bedrock" as used in soil survey refers to continuous, coherent (consolidated) rock. It can be a physical barrier within the solum that limits rooting depth or the immediate parent material source for residual soils. Bedrock helps to determine local topography and the soils that form across it. It can also indirectly impact soils. If fairly close to the base of the solum, bedrock can affect the presence or absence of ground water and preferential flow direction, depending upon its porosity. It can be a determinant factor in slope stability (tendency for landslides) or impact excavation, such as soil suitability for basements and road construction. Identifying the bedrock, whether its influence is direct or indirect, is essential in understanding the intrinsic chemical and physical behavior of both the rock material itself and its soil or regolith derivatives. Bedrock also has a major impact on soil geography and the accurate prediction of it. Boundaries between types of bedrock commonly, but not always, coincide with changes in overlying soil types. Therefore, accurate recognition and documentation of bedrock is generally essential. In some natural settings, the documentation of bedrock may be problematic, impractical, or unnecessary. Bedrock is not recorded if it does not exert substantial influence on the soil. An example is bedrock that is deeply buried by regolith, such as till, basin fills, and coastal or lacustrine sediments.

Geological materials need to be defined in accordance with the accepted standards and nomenclature of geology. The accepted, authoritative names of the geological formations are recorded in soil descriptions. As soil research progresses, there is an increasing number of correlations between particular geological formations and the mineral and nutrient content of parent materials and soils. Examples include:

- Certain terrace materials and deposits of volcanic ash that are different in age or source, but otherwise indistinguishable, may vary widely in content of cobalt.
- The phosphorus content of otherwise similar soils may vary widely due to similar limestones that can be distinguished in the field only by specific fossils.

Igneous rocks formed by the solidification of magma that originated within Earth's upper mantle. There are two main types based on their mode of formation—intrusive and extrusive. Intrusive (syn., plutonic) types form at considerable depth in the Earth's crust and possess a coarse grain texture due to the slow cooling of magma. Examples of intrusive igneous rocks that weather to soil parent material are granite, diorite, and gabbro. Extrusive (syn., volcanic) types form on the Earth's surface or at very shallow depth and possess a fine grain texture due to the rapid cooling of magma. Examples of common extrusive igneous rocks are rhyolite, andesite, and basalt.

Sedimentary rocks formed from sediments laid down in previous geological ages. The principal broad groups of sedimentary rocks are clastic, chemical, and organic. Examples of rock lithologies in the clastic group are shale, sandstone, and conglomerate; examples of those in the chemical group are limestone, gypsum rock, and travertine; and examples of those in the organic group are coal and diatomite. There are many varieties of these lithologies. For example, chalk is a soft variety of limestone. Many lithologies are intermediate between the broad groups. Examples are calcareous sandstone and arenaceous limestone.

Metamorphic rocks resulted from profound alteration of igneous and sedimentary rocks by heat and pressure. General classes of metamorphic rocks important as parent material are gneiss, schist, slate, marble, quartzite, and phyllite.

Kinds of Bedrock

Kind is the most important bedrock feature to describe. It indicates the general composition of the rock and how the rock and its weathering products are likely to behave. Soil survey follows standard conventions for rock type compositions and names. There is a large number of officially recognized rock types. They are very detailed and can be functionally cumbersome for soil survey. Subsequently, soil survey tends to focus on broader categories and common rock types, particularly those found in the near surface environment. More obscure or minor rock types can be recognized if they have important economic or environmental impact. Some databases maintain long alphabetical master lists of bedrock. A helpful way to arrange the large number and variety of bedrock kinds is to separate them into widely recognized subsets, such as igneous, metamorphic, and sedimentary. In addition to bedrock kind, descriptions of bedrock should include information about the spacing of fractures, degree of weathering, and depth to contact (if within or near the solum).

The general groups of bedrock types described earlier can be subdivided or rearranged slightly to provide groups of bedrock types commonly pertinent to soils:

- Igneous-intrusive: Examples are anorthrosite, diabase, and granite.
- Igneous-extrusive: Examples are a'a lava, andesite, and basalt.
- Igneous-pyroclastic: Examples are pyroclastic flow, tuff, and volcanic breccia.
- Metamorphic: Examples are amphibolite, gneiss, and schist.
- Sedimentary-clastics: Examples are arenite, argillite, and mudstone.
- Interbedded: Examples are limestone-sandstone, sandstone-shale, and shale-siltstone.
- Evaporites, organics, and precipitates: Examples are tufa, coal, and limestone.

Depth to Bedrock

Depth to bedrock is a crucial feature because of its impact on plant growth, internal water dynamics and direction, and land management. This is particularly true for agricultural soils if hard bedrock is within 2 meters of the surface. Bedrock can limit rooting depth, reduce the potential soil water supply, affect internal water flow, and impact various mechanical activities, such as deep ripping, foundation excavation, fence post placement, and suitability for basements. The depth from the ground surface to the contact with bedrock should be recorded.

Fracture Interval

Most bedrock contains a natural joint or crack network that functions as by-pass flow routes for internal water. These fractures can vary substantially in the amount of water they are able to transmit and can potentially affect pond integrity, internal pollutant movement, and water well yields. If observable, the average horizontal spacing between vertical rock joints in the bedrock layer is described.

Weathering

Not all bedrock is chemically and/or physically altered from its pristine state to the same extent. Weathering generally increases porosity and the water-holding capacity and reduces bulk density and coherency. A weathering class (e.g., slight, moderate, strong) can be assigned to record the bedrock's subjective extent of weathering as compared to its presumed un-weathered state.

Lithostratigraphic Units

Lithostratigraphic units are mappable rock or sediment bodies. In geochronology, younger units overly older units (law of superposition). Regolith units, both unconsolidated material and bedrock units, are identified and named according to standard conventions of the International Stratigraphic Code. Table lists these units in descending rank. This naming system provides a standard, shorthand method of identifying and concisely communicating information on strata and rock type. It aids recognition of differences in geology and the soils developed in or on them. Some soils, particularly residual soils, can be linked to specific bedrock units. Other soils, such as loess, can occur across multiple bedrock units if their lithostratigraphic unit is not bedrock constrained. If possible, the hierarchical lithostratigraphic units at a site should be recorded.

Lithostratigraphic units and their hierarchical rank and definition:

- Supergroup: The broadest lithostratigraphic unit. A supergroup is an assemblage of related, superposed groups, or groups and formations. It is most useful for regional synthesis.
- Group: The second ranking lithostratigraphic unit. A group is a named assemblage of superposed formations and may include unnamed formations. It is useful for small-scale (broad) mapping and regional stratigraphic analysis.
- Formation (or Geologic Formation): The basic lithostratigraphic unit used to describe, delimit, and interpret sedimentary, extrusive igneous, metavolcanic, and metasedimentary rock bodies (excluding metamorphic and intrusive igneous rocks). It is based on lithic characteristics and stratigraphic position. A formation is commonly, but not necessarily, tabular and stratified and is of sufficient extent to be mappable at the Earth's surface or traceable in the subsurface at conventional mapping scales.
- Member: The formal lithostratigraphic unit next in rank below a formation and always part of a formation. A formation need not be divided selectively or entirely into members. A member may extend laterally from one formation to another.
- Lens (or Lentil): A specific type of member. A lens is a geographically restricted member that terminates on all sides within a formation.
- Tongue: A specific type of member. A tongue is a wedge-shaped member that extends beyond the main formation boundary or that wedges or pinches out within another formation.
- Bed: The smallest lithostratigraphic unit of sedimentary rock. A bed is a subdivision of a member based upon distinctive characteristics or economic value (e.g., coal member). Members need not be divided selectively or entirely into beds.
- Flow: The smallest lithostratigraphic unit of volcanic rock. A flow is a discrete, extrusive, volcanic body distinguishable by texture, composition, super-position, and other criteria.

Erosion

Erosion is the detachment and movement of soil material. The process may be natural or accelerated by human activity. Depending on the local landscape and weather conditions, erosion can range from very slow to very rapid. Loss of the soil surface layer has a direct detrimental impact on site productivity and on off-site sedimentation and nutrient inputs. It is especially important to evaluate for environmental and agronomic purposes. The dominant kind and degree (relative magnitude) of accelerated erosion at the site should be estimated.

Natural Erosion

Naturally occurring erosion sculptured landforms on the uplands and built landforms on the lowlands. Its rate and distribution in time control the age of land surfaces and many of the internal properties of the soils on them. The formation of the Channel Scablands in the State of Washington is an example of extremely rapid natural or geologic, erosion. The broad, nearly level inter stream divides on the Coastal Plain of the south-eastern United States are examples of areas with very slow or no natural erosion.

Landscapes and their soils are evaluated from the perspective of their natural erosional history. Evidence that material has been moved and re-deposited, including buried soils, stone lines, and deposits of windblown material, is helpful in understanding natural erosion history. Thick weathered zones that developed under earlier climatic conditions may have been exposed and become the material in which new soils formed. In landscapes of the most recently glaciated areas, the consequences of natural erosion, or lack of it, are less obvious than where the surface and the landscape are early Pleistocene or even Tertiary in age. However, even on the landscapes of the most recent glaciation, postglacial natural erosion may have redistributed soil materials on the local landscape. Natural erosion is an important process that affects soil formation and, like human-induced erosion, can remove all or part of soils formed in the natural landscape.

Accelerated Erosion

Accelerated erosion is largely the consequence of human activities, primarily those that result in a loss of soil cover, such as tillage, grazing, and cutting of timber. Kinds are listed in table and discussed below:

Table: Kinds of Accelerated Erosion.

Erosion kind	Criteria
Wind	Deflation by wind
Water:	Removal by running water
Sheet	Relatively uniform soil loss; no channels
Rill	Small channels (can be obliterated by conventional tillage)
Gully	Big channels (cannot be obliterated by conventional tillage)
Tunnel	Subsurface voids within soil that are enlarged by running water (i.e., piping)

The rate of erosion can be increased by events besides human activities. For example, fire that destroys vegetation can trigger erosion. Spectacular episodes of erosion, such as the soil blowing on the Great Plains of the central United States in the 1930s, have not all been due to human activities; frequent dust storms were recorded on the Great Plains before the region became a grain-producing area.

Accelerated erosion may not be easy to distinguish from natural erosion on some soils. A distinction can be made by studying and understanding the sequence of sediments and surfaces on the local landscape as well as by studying soil properties. For example, in some areas of the eastern

United States, native forests were cut and burned to create cropland. In some places where the soils were particularly susceptible, this resulted in extensive soil erosion. The sediments from the uplands can be observed on adjacent flood plains as a sequence of layers that, in some places, are up to a few meters thick over a buried soil. The contact between the original soil surface and new sediments commonly is evidenced by numerous pieces of charcoal above the contact, which presumably originated from the burning of timber.

Wind Erosion

The term "wind erosion," as used in this manual, in soil science generally, and by many geologists, indicates the detachment, transportation, and deposition of soil particles by wind, not the sculpture of rocks by windblown particles. Wind erosion in regions of low rainfall can be widespread, especially during periods of drought. Unlike water erosion, wind erosion is generally not related to slope gradient. The hazard of wind erosion is increased by removing or reducing the amount of vegetation. When winds are strong, coarser particles are rolled or swept along, on or near the soil surface and finer particles are forced into the air. The particles are deposited in places sheltered from the wind. When wind erosion is severe, the sand particles may drift back and forth locally with changes in wind direction while silt and clay are carried away. Small areas in which the surface layer has blown away may be associated with areas of deposition in such an intricate pattern that the two cannot be identified separately on soil maps.

Water Erosion

Water erosion results from the removal of soil material by flowing water, including the detachment of soil material by the impact of raindrops. The soil material is suspended in runoff water and carried away. Some sediment may be carried just a few meters before being deposited, while other sediment may be completely removed from the site. Four kinds of accelerated water erosion are commonly recognized: sheet, rill, gully, and tunnel (piping).

Sheet erosion is the more or less uniform removal of soil from an area without the development of conspicuous water channels. The channels are tiny or tortuous, exceedingly numerous, and unstable. They enlarge and straighten as the volume of runoff increases. Sheet erosion is less apparent, particularly in its early stages, than other types of erosion. It can be a problem for soils that have a slope gradient of only 1 or 2 percent; however, it is generally more of an issue as slope gradient increases.

Rill erosion is the removal of soil as concentrated runoff cuts many small, but conspicuous, channels. It is intermediate in degree between sheet and gully erosion. The channels are shallow enough that they are easily obliterated by tillage. After an eroded field has been cultivated, determining whether soil losses resulted from sheet or rill erosion is generally impossible.

Gully erosion is the removal of soil by water along the line of flow. Gullies form in exposed natural drainage ways, in plow furrows, in animal trails, in vehicle ruts, between rows of crop plants, and below broken human-made terraces. Unlike rills, they cannot be obliterated by ordinary tillage. Deep gullies cannot be crossed with common types of farm equipment.

Gullies and gully patterns vary widely. V-shaped gullies form in material that is equally or increasingly resistant to erosion with increasing depth. U-shaped gullies form in material that is equally or decreasingly resistant to erosion with depth. As the substratum is washed away, the overlying material loses its support, falls into the gully, and is also washed away. Most U-shaped gullies become modified toward a V shape once the channel stabilizes and the banks begin to crumble and slump. The maximum depth to which gullies are cut is governed by resistant layers in the soil, by bedrock, or by the local base level. Many gullies develop headward, i.e., they extend up the slope as the gully deepens in the lower part.

Tunnel erosion may occur in soils with subsurface horizons or layers that are more subject to entrainment than the surface horizon or layer. Through ponded infiltration, the free water enters into the soil's surface-connected macropores. Desiccation cracks and rodent burrows are examples of macropores that may initiate the process. The soil material incorporated into the moving water travels downward within the soil profile, and if there is an outlet, may move out of it completely. As a result, tunnels (also referred to as pipes) form, enlarge, and coalesce. The portion of the tunnel near the inlet may enlarge disproportionately to form a funnel-shaped feature, commonly referred to as a "jug." This phenomenon is called "piping" or "jugging" and occurs especially in areas with appreciable amounts of exchangeable sodium.

Sediment carried by water typically is deposited wherever the water's velocity slows, such as at the mouth of gullies, at the base of slopes, along stream banks, on alluvial plains, in reservoirs, and at the mouth of streams. Water moving rapidly can deposit stones. As it slows, it deposits cobbles, followed by gravel, sand, and finally silt and clay. The slope length for sediment transport is the distance from the highest point on the slope where runoff may start to where the sediment in the runoff would be deposited.

Estimating the Degree of Erosion

Soil examinations can estimate the degree to which accelerated erosion has modified the soil. However, estimating the amount of surface soil that is no longer present can be very difficult. This is generally most feasible if sufficient areas of the soil are known to be little affected by past accelerated erosion and can be used for comparison studies. The recognition of eroded and uneroded phases of a soil is useful if at least some soil properties making up the eroded phase are different enough from those of the uneroded phase to impact the soil's use and management. The eroded soil is identified and classified on the basis of the properties of the soil that remains and not on what was presumed to have been present in the past. In some cases, the eroded soil may classify differently from the uneroded soil. An estimate of the soil lost is described. Eroded soils are defined so that the boundaries on the soil maps separate soil areas with different use suitabilities and different management needs.

The depth to a reference horizon or soil characteristic in areas under a use that has minimized erosion are compared to the same properties in areas under uses that have accelerated erosion. For example, a soil that supports native grass or large trees with no evidence of cultivation could be compared with the same or similar soil that has been cleared and cultivated for a relatively long time. The depth to reference layers is measured from the top of the mineral soil because cultivation destroys organic horizons at the surface.

The depth to a reference layer must be interpreted in terms of recent soil use or history. The upper parts of many forested soils have roots that make up as much as one-half of the soil volume. When these roots decay, the soil settles. Removal of rock fragments can also lower the surface.

Cultivation may cause differences in thickness of layers. The thickness of surficial zones that have been bulked by tillage should be adjusted downward to what they would be under natural conditions.

The thickness of a plowed layer cannot be used as a standard for either losses or additions of material because, as a soil erodes, the plow cuts progressively deeper. Nor can the thickness of the uncultivated and uneroded A horizon be used as a standard for all cultivated soils, unless the A horizon is much thicker than the plow layer. If the horizon immediately below the plowed layer of an uneroded soil is distinctly higher in clay than the A horizon, the plow layer becomes progressively more clayey under continued cultivation as erosion progresses. In this case, the texture of the plow layer can be a criterion of erosion. Comparisons must be made on comparable slopes. Near the upper limit of a soil's range of slope gradient, horizons may normally be thinner than near the lower limit.

Roadsides, cemeteries, fence rows, and similar uncultivated areas that make up a small part of the landscape or were subject to unusual cultural histories must be used cautiously for setting standards. In these areas, the reference standards for surface layer thickness are generally set too high. In naturally treeless areas or in areas cleared of trees, dust may collect in fence rows, along roadsides, and in other small uncultivated areas that are covered with grass or other stabilizing plants. This accumulated dust may cause the surface horizon to become several centimeters thicker in a short time.

For soils having clearly defined horizons, differences due to erosion can be accurately determined by comparison to undisturbed or uncultivated sites. Guidelines for estimating erosion for soils with a thin A horizon and little or no other horizon are more difficult to establish. After the thin surface layer is gone or has been mixed with underlying material, few clues remain for estimating the degree of erosion. One must rely on the physical conditions of the material in the plowed layer, the appearance and amount of rock fragments on the surface, the number and shape of gullies, and similar evidence. For many soils having almost no horizon expression, attempting to estimate the degree of erosion serves no useful purpose.

Precise estimates of the amount of soil lost from a site based on comparison studies with a similar uneroded site are complicated by several factors. The goal is to establish map unit concepts that reflect the relative degrees of soil loss between eroded phases of a soil and that result in some significant differences in the use and management of the soils based on their current properties.

Degree Classes for Accelerated Erosion

The degree classes for accelerated erosion discussed below and listed in table apply to both water and wind erosion. They are not applicable to landslip or tunnel erosion. The classes pertain to the proportion of upper horizons that has been removed. These horizons may range widely in thickness; therefore, the absolute amount of erosion is not specified.

Degree class	Criteria: Estimated % loss of the original combined A + E horizons, or the estimated loss of the upper 20 cm (if original, combined A + E horizons are < 20 cm thick)
None	0 %
1	> 0 to 25 %
2	25 to 75 %
3	75 to 100 %
4	> 75 % and total removal of the A horizon

1. Class 1: This class consists of soils that have lost some, but on average less than 25 percent, of the original A and/or E horizons or of the uppermost 20 cm if the original A and/or E horizons were less than 20 cm thick. Throughout most of the area, the thickness of the surface layer is within the normal range of variability of the uneroded soil. Less than 20 percent may consist of scattered small areas with a significantly modified surface layer.

Evidence for class 1 erosion includes:

- A few rills.
- Accumulation of sediment at the base of slopes or in depressions.
- Scattered small areas where the plow layer contains material from below.
- Evidence of the formation of widely spaced, deep rills or shallow gullies without consistently measurable reduction in thickness or other change in soil properties between the rills or gullies.

2. Class 2: This class consists of soils that have lost, on average, 25 to 75 percent of the original A and/or E horizons or of the uppermost 20 cm if the original A and/or E horizons were less than 20 cm thick. Throughout most cultivated areas of class 2 erosion, the surface layer consists of a mixture of the original A and/or E horizons and material from below. Some areas may have intricate patterns, ranging from somewhat over-thickened surface layers where sediment has accumulated locally to small areas of uneroded soils on gentle slopes or severely eroded soils on steeper, convex slopes. Where the original A and/or E horizons were very thick, little or no mixing of underlying material may have taken place.

3. Class 3: This class consists of soils that have lost, on average, 75 percent or more of the original A and/or E horizons or of the uppermost 20 cm if the original A and/or E horizons were less than 20 cm thick. In most areas, material below the original A and/or E horizons is exposed at the surface, especially in convex positions in cultivated areas; the plow layer consists entirely or largely of this material. Even where the original A and/or E horizons were very thick, at least some mixing with underlying material generally took place. Despite the generally universal loss of surface soil, some areas exhibit intricate patterns, ranging from somewhat over-thickened surface layers where sediment has accumulated locally to small areas of only slightly eroded soils, generally where slopes are relatively gentle.

4. Class 4: This class consists of soils that have lost all of the original A and/or E horizons or the uppermost 20 cm if the original A and/or E horizons were less than 20 cm thick. In most areas, some or all of the deeper horizons have been removed throughout the majority of the area. The original soil can be identified only in small areas. Some areas may be smooth, but most have an intricate pattern of gullies.

Land Cover

The type of land cover around the site where a soil is described should be recorded. As with other descriptive terms, it is best to use standard terms consistently. The NCSS has adopted a set of general terms that includes land cover kinds, such as artificial cover, barren land, crop cover, shrub cover, grass/herbaceous cover, tree cover, and water. Subtypes within these general classes are also recorded.

In addition to recording the overall land cover condition at the site, more detailed analysis can be done to provide quantitative estimates of the surface cover at the site. The ground surface of most soils is covered by vegetation to some extent at least part of the year. In addition, rock fragments form part of the mineral material at the surface of many soils. The vegetal material that is not part of the surface horizon and the rock fragments together form the ground surface cover. The proportion of cover, along with its characteristics, is very important in determining a soil's thermal properties and resistance to erosion.

At one extreme, estimation of cover can be made visually without quantitative measurement. At the other extreme, transect techniques can be used to make an almost complete modal analyses of the ground surface. If the ground surface is relatively permanent, more effort in documentation is justified. In many cases, a combination of rapid visual estimates and transect techniques is appropriate.

The ground surface may be divided into fine earth and material other than fine earth. The latter consists of rock fragments and both live and dead vegetation. Vegetation is separated into canopy and noncanopy (litter). A canopy component has a relatively large cross-sectional area capable of intercepting rainfall compared to the area near enough to the ground surface to affect overland water flow. When determining susceptibility to erosion, both canopy and noncanopy vegetation are considered.

The first step in evaluation is determining the components (typically one to three) of the ground surface cover. A common three-component land surface consists of trees, bushes, and areas between the two. The areal proportion of each component must be established, such as by transect. If a canopy component is present, the area within the tree drip line (edge of where water drips from trees and onto the ground) is determined as a percent of the ground surface. For each canopy component, the effectiveness must be established. Effectiveness is the percent of vertical raindrops that would be intercepted. The canopy effectiveness is typically estimated visually, but a spherical densitometer may be used. In addition to the canopy effectiveness, the mulch must be identified for each component.

Transect techniques may be used to determine the mulch percentage. The mulch can be subdivided into rock fragments and vegetation. From the areal proportions of the components and their respective canopy efficiencies and mulch percentages, the soil-loss ratio may be computed for the whole land surface. Other observations may include the percent of kinds of plants, size of rock fragments, amount of green leaf area, and aspects of color of the immediate surface that may affect absorption of radiant energy in an area.

Vegetation

It is important to evaluate and record details about the vegetative community of a site, particularly in non-agricultural settings, such as rangeland, marshes, and forests. Plants reflect the integrated effects of controlling water dynamics, climate, native fertility, human intervention, and other factors upon the soil. Baseline vegetation information centers on the plant species present and the extent of the area that they cover.

Typically, the dominant plant species present are identified and documented in descending order of prominence. The scientific name is used along with, or in lieu of, the common name. Common plant names are not preferred as they may not be unique. A species may be known by multiple common names in a region, depending upon local cultures and languages spoken. The appropriate scientific plant symbol is also recorded, for example, ANGE (Andropogon gerardii, or big bluestem). The amount of ground covered by each plant species recorded at the site is also estimated or measured.

References

- What-geomorphology: geomorphology.org.uk, Retrieved 11, July 2020
- Biogeomorphology-of-coastal-systems: coastalwiki.org, Retrieved 26, February 2020

Chapter 2

Concepts and Models in Geomorphology

Models in geomorphology are used for the conceptual and quantitative representations of the surface of the Earth. There are various models which are used in geomorphology such as hardware models, mathematical model and experimental design models. This chapter has been carefully written to provide an easy understanding of these models in geomorphology.

Like any other discipline Model building in Geomorphology plays a very significant role, this was very vividly highlighted by various geographers. The progression of abridging real landscapes to manageable proportions is model building. Defined in a general way, a geomorphic model is a simplified representation of some aspect of a real landscape that happens to interest a geomorphologist. It is an attempt to describe, analyse, simplify or display a geomorphic system. Model building in geomorphology, has been strongly influenced by the systems analysis and system approach is most suitable to provide insights and make predictions. The dynamics of different systems are explained by these geomorphic models and the major construed systems in geomorphology.

General Systems Theory

The general system approach to the study of landforms rests upon some extensive attitude to groups of geomorphic phenomena which is obtained as the consequence of experience or insight. For this reason, the prominence lies in the organisation and operation of the system as an entire system or as linked components, to a certain extent, rather than in detail study of individual system elements.

A geomorphic system is an inclusive composite of landforms which operate together according to some apparent pattern, energy and matter input into the system giving rise to a predictable system response in terms of internal organization and the consequential energy and matter output. Geomorphic systems can be regarded as part of the super systems and as being comprised of subsystems. Subsystems are consequently the indispensable components of a system and can be acknowledged as distinct input-output linkages.

In geomorphology subsystems are generally united by pouring the output of one of the subsystems into an additional subsystem to form its input. When interest in the subsystem operations is extremely detailed, the systems approach is substituted by the experimental design model. Two systems are said to have 'similarity' when there is precise equality of all components, and to have equivalence when they transform the same inputs in to indistinguishable outputs. Synthetic system, partial system and black box are components of general system and dealt with in detail.

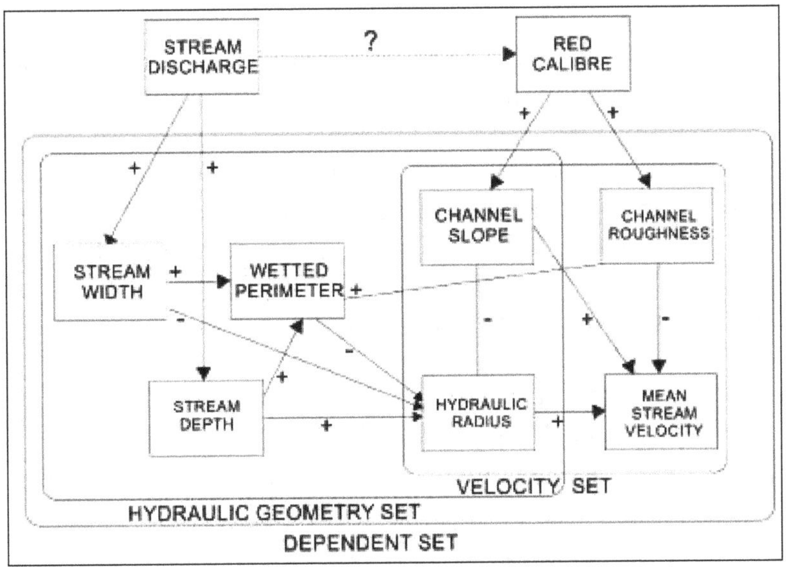

General System – Hydraulic Geometry.

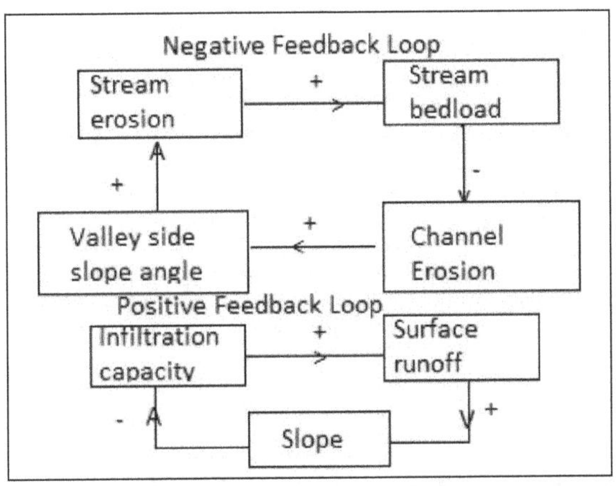

Feedback Relation in System.

Synthetic Systems

Synthetic systems in their preliminary phases exhibit similarity with experimental design models, being involved with recognizing allegedly critical features of the structure of geomorphic phenomena, together with their sampling and analysis. The purpose of synthetic systems research, on the other hand, goes further than this in that its objective lies in the synthesis of the analysed structure and the extension and generalization of conclusions to the point of making process-response models.

The modus operandi for building up synthetic systems starts with the recognition of definite vital structural elements in a given geomorphic complex, together conceivably with some views as to the potential relationship between these elements (involving direction of causation and which variables change together).

Partial Systems

The partial systems are one which have been first and foremost associated with the resolving of fundamentally practical problems in the earth sciences. It has been related to the establishment of workable relationships between often randomly grouped sets of factors or subsystems. Detailed knowledge on the subject of the internal functioning of these subsystems is not regarded as obligatory, but the attainment of precise information about the interrelationships between these subsystems allows one to recognize and forecast the behavior of the entire systems under different input conditions.This information may come from diverse sources, but the object is always to set up mathematical relationships between the system inputs and outputs, without any effort to illustrate unambiguously the internal mechanisms of the system. The consequential behavior pattern are then used to envisage, and to a smaller extent to comprehend, the behavior patterns in other similar systems or in same system at different times.

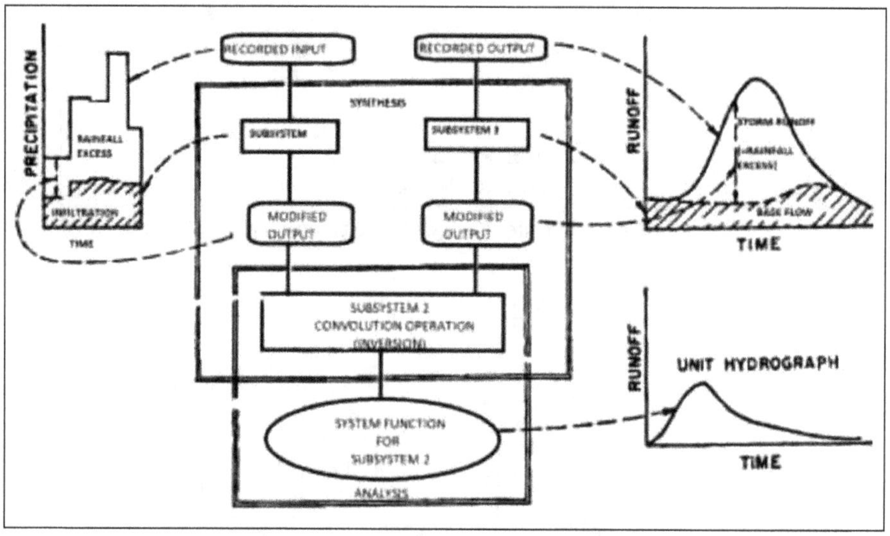

Partial System

The Black Box

On the whole intense application of general systems models engages the concept of the 'black box'. This necessitates modest or no detailed information concerning the system components or subsystems within the black box system, interest being focused upon the nature of the outputs which result from differing outputs. Consequently, the black box is equivalent to the grey box of the partial system and the 'white box' of the synthetic system.

In geomorphology, the black box approach has supported much of the significant work of the last one hundred years. Such work has been characterized by broad perceptive leaps wherein decisions are taken on the subject of the supposed dominant characteristics of broad landform assemblages. Even though there has been much guesswork and assumption on the subject of the detailed nature of the progression at work, the characteristic patterns of landform transformations through time, and the rates of operation of geomorphic processes.

During the past decade several valuable attempts have been made to apply general systems theory to the study of geomorphology, with a view to examining in detail the fundamental basis of the subject, its aims and its methods. They come at a time when the conventional approach is in danger of subsiding into an uncritical series of conditioned reflexes, and when the more imaginative modern work in geomorphology often seems to be sacrificing breadth of vision for focus on details. In both approaches it is a common trend for workers to be increasingly critical of operating within general frameworks of thought, particularly with the examples of the Davis and Penck geormorphic systems before them, and "classical" geomorphologists have retreated into restricted historical studies of regional form elements, whereas, similarly, quantitative workers have often withdrawn into restricted empirical and theoretical studies based on process.

It is wrong, however, to confuse the restrictions which are rightly associated with preconceived notions in geomorphology with the advantages of operating within an appropriate general systematic framework. The first lead to the closing of vistas and the decrease of opportunity; the second, however, may increase the scope of the study, make possible correlations and associations which would otherwise be impossible, generally liberalize the whole approach to the subject and, in addition, allow integration into a wider general conceptual framework. Essentially, it is not possible to enter into a study of the physical world without such a fundamental basis for the investigation, and even the most qualitative approaches to the subject show very strong evidence of operations of thought within a logical general framework, albeit a framework of thought which is in a sense unconscious. For example, has pointed to the essential difference between the approaches to geomorphology of Gilbert and Davis, and in this respect the fundamental value of the adoption of a suitable general framework of investigation based on general systems theory becomes readily apparent.

Following the terminology it is possible to recognize in general two separate systematic frameworks wherein one may view the natural occurrence of physical phenomena; the closed system and the open. Hall and Fagan have defined a system as a set of objects together with relationships between the objects and between their attributes." In the light of this definition, it is very significant that one of the fundamental purposes of Davis' approach to landforms was to study them as an assemblage, in which the various parts might be related in an area and a time sense, such that different systems might be compared, and the same system followed through its sequence of time changes. Closed systems are those which possess clearly defined closed boundaries, across which no import or export of materials or energy occurs. This view of systems immediately precludes a large number, perhaps all, of the systems with which natural scientists are concerned; and certainly most geographical systems are excluded on this basis, for boundary problems and the problems of the association between areal units and their interrelationships lie very close to the core of geographical investigations.

Another characteristic of closed systems is that, with a given amount of initial free, or potential, energy within the system, they develop toward states with maximum "entropy". Entropy is an expression for the degree to which energy has become unable to perform work. The increase of entropy implies a trend toward minimum free energy. Hence, in a closed system there is a tendency for levelling down of existing differentiation within the system; or, for progressive degradation of energy into its lowest form, i.e. heat as undirected molecular movement. This is expressed by the second law of thermodynamics which, in its classic form, is formulated for closed systems. In such systems, therefore, the change of entropy is always positive, associated with a decrease in the amount of free energy, or, to state this another way, with a tendency toward progressive destruction of existing order or differentiation.

Thus, one can see that Davis' view of landscape development contains certain elements of closed system thinking including, for example, the idea that uplift provides initially a given amount of potential energy and that, as degradation proceeds, the energy of the system decreases until at the stage of peneplanation there is a minimum amount of free energy as a result of the levelling down of topographic differences. The Davisian peneplain, therefore, may be considered as logically homologous to the condition of maximum entropy, general energy properties being more or less uniformly distributed throughout the system and with a potential energy approaching zero. The positive change of entropy, and connected negative change of free energy, implies the irreversibility of events within closed systems. This again bears striking similarities to the general operation of the geomorphic cycle of Davis. The belief in the sequential development of landforms, involving the progressive and irreversible evolution of almost every facet of landscape geometry, in sympathy with the reduction of relief, including valley-side slopes and drainage systems, is in accord with closed system thinking. Although "complications of the geographical cycle" can, in a sense, put the clock back, nothing was considered by Davis as capable of reversing the clock. The putting back of the clock by uplift, therefore, came to be associated with a release, or an absorption into the new closed system, of an increment of free energy, subsequently to be progressively dissipated through degradation.

Also, in closed systems there is the inherent characteristic that the initial system conditions, particularly the energy conditions, are sufficient to determine its ultimate equilibrium condition. This inevitability of closed-system thinking is very much associated with the view of geomorphic change held by Davis. Not only this but the condition of a closed system at any particular time can be considered largely as a function of the initial system conditions and the amount of time which has subsequently elasped. Thus closed systems are eminently susceptible to study on a time, or historical, basis. This again enables one to draw striking analogies between closed-system thinking and the historical approach to landform study which was proposed by Davis.

Finally, it is recognized that closed systems can reach a state of equilibrium. Generally speaking, however, this equilibrium state is associated with the condition of maximum entropy which cannot occur until the system has run through its sequential development. In addition, it is impossible to introduce the concept of equilibrium into a closed-system framework of thought without the implication that it is associated with stationary conditions. The only feature of the cyclic system of Davis which employed the general concept of equilibrium was that of the "graded" condition of stream channels and slopes which, significantly, Davis borrowed from the work of Gilbert, who had an entirely noncyclic view of landform development. Characteristically, the concept of grade was the one feature of Davis' synthesis which seems least well at home in the cyclic framework, for it has always proved difficult to imagine how, within a closed system context, a graded or equilibrium state could exist and yet the associated forms be susceptible to continued change namely, down cutting or reduction.

The foregoing is not meant to imply that it is unprofitable to consider any assemblage of phenomena within a closed system framework, or, as Davis did, to overstress those aspects or phases which seem to achieve most significance with reference to the closed system model. It is important, however, to recognize the sources of partiality which result, not from any inherent quality of the data itself, but from the general systematic theory under which one is operating. In reality, no systematic model can encompass the whole of a natural complex without ceasing to be a model and the phenomena of geomorphology present problems both when they are viewed within closed and open systematic frameworks. In the former, the useful concept of dynamic equilibrium or grade rests most uncomfortably in the latter, as will be seen, the progressive, loss of a component of potential energy due to relief reduction imposes an unwelcome historical parameter.

A simple, classic example of a closed system is represented by a mass of gas within a completely sealed and insulated container. If, initially, the gas at one end of the container is at a higher temperature than that at the other, this can be viewed as a condition of maximum segregation, maximum free energy, and, consequently, of maximum ability to perform work, should this thermal gradient be harnessed within a larger closed system. This is the state of minimum entropy. It is obvious however that this state of affairs is of a most transient character and that immediately an irreversible heat flow will begin toward the cooler end of the container. This will progressively decrease the segregation of mass and energy within the system, together with the available free energy and the 'ability of this energy to perform work, bringing about a similarly progressive increase of entropy. While the system remains closed nothing can check or hinder this inevitable levelling down of differences, which is so predictable that, knowing the initial energy conditions, the thermal conductivity of the gas and the lapse of time, one could accurately calculate the thermal state of the system at any required stage. Thus the distribution of heat energy and the heat flow within the system have a progressive and sequential history, the one becoming less segregated and the other ever-decreasing. Nor is it possible to imagine any form of equilibrium until all the gas has attained the same temperature, when the motion of the gas molecules is quite random and the static condition of maximum entropy obtains.

Open systems contrast quite strikingly with closed systems. An open system needs an energy supply for its maintenance and preservation and is in effect maintained by a constant supply and removal of material and energy. Thus, direct analogies exist between the classic open systems and drainage basins, slope elements, stream segments and all the other for mass emblages of a landscape. The concept of the open system includes closed systems, however, because the latter can be considered a special case of the former when transport of matter and energy into and from the system becomes zero. An open system manifests one important property which is denied to the closed system. It may attain a "steady state" wherein the import and export of energy and material are equated by means of an adjustment of the form, or geometry, of the system itself. It is more difficult to present a simple mechanical analog to illustrate completely the character and operations of an open system but it may be helpful to visualize one such system as represented by the moving body of water contained in a bowl which is being constantly filled from an overhead inflow and drained by an outflow in the bottom. If the inflow is stopped, the bowl drains and the system ceases to exist; whereas, if the inflow is stopped and the outflow is blocked, the system partakes of many of the features of a closed system. In such an arrangement, changes in the supply of mass and energy from outside lead to a self-adjustment of the system to accommodate these changes. Thus, if the inflow is increased, the water level in the basin rises, the head of water above the outflow increases, and the outflow discharge will increase until it balances the increased inflow. At this time the level of water in the bowl will again become steady.

Long ago, Gilbert recognized the importance of the application of this principle of self-adjustment to landform development. The tendency to equilibrium of action, or to the establishment of a dynamic equilibrium, has already been pointed out in the discussion of the principles of erosion and of sculpture, but one of its most important results has not been noticed.

Of the main conditions which determine the rate of erosion, namely, the quantity of running water, vegetation, texture of rock, and declivity, only the last is reciprocally determined by rate of erosion. Declivity originates in upheaval, or in the displacement of the earth's crust by which mountains and continents are formed: but it receives its distribution in detail in accordance with the laws of erosion. Wherever by reason of change in any of the conditions the erosive agents come to have locally exceptional power, that power is steadily diminished by the reaction of the rate of erosion upon declivity. Every slope is a member of a series, receiving the water and the waste of the slope above it, and discharging its own water and waste upon the slope below. If one member of the series is eroded with exceptional rapidity, two things immediately result: first, the member above has its own level of discharge lowered, and its rate of erosion is thereby increased; and second, the member below, being clogged by an exceptional load of detritus, has its rate of erosion diminished. The acceleration above and the retardation below diminish the declivity of the member in which the disturbance originated: and as the declivity is reduced, the rate of erosion is likewise reduced.

But the effect does not stop here. The disturbance that has been transferred from one member of the series to the two which adjoin it, is by then transmitted to others, and does not cease until it has reached the confines of the drainage basin. For in each basin all lines of drainage unite in a main line, and a disturbance upon any line is communicated through it to the main line and thence to every tributary. And as a member of the system may influence all the others, so each member is influenced by every other. There is interdependence throughout the system.

This form-adjustment is brought about by the ability of an open system for self-regulation. Le Chatelier's Principle (originally stated for equilibrium in closed systems) can be expanded also to include the so-called "Dynamic Equilibrium" or steady states in open systems.

Any system in equilibrium undergoes, as a result of a variation in one of the factors governing the equilibrium, a compensating change in a direction such that, had this change occurred alone it would have produced a variation of the factor considered in the opposite direction.

A geomorphic statement of this principle has been given: A graded stream is one in which, over a period of years, slope is delicately adjusted to provide, with available discharge and with prevailing channel characteristics, just the velocity required for the transportation of the load supplied from the drainage basin. The graded stream is a system in equilibrium; its diagnostic characteristic is that any change in any of the controlling factors will cause a displacement of the equilibrium in a direction that will tend to absorb the effect of the change.

The cyclic adaptation of the concept of grade did not give sufficient importance to the factors, other than channel slope, which a stream system can control for itself, and in this respect Davis' ignorance of the significance of the practical experiments of Gilbert is most evident. A stream system cannot greatly control its discharge, which represents the energy and mass which is externally supplied into the open system. Neither can it completely control the amount and character of the debris supplied to it, except by its action of abrasion and sorting or as the result of the rapport which seems to exist regionally between stream-channel slope and valley-side slope. However, besides adjusting the general slope of its channel by erosion and deposition, a stream can very effectively and almost instantaneously control its transverse channel characteristics, together with its efficiency for the transport of water and load, by changes in depth and width of the channel.

The downstream curves on Brandywine Creek suggest that the adjustment of channel shape may be as significant as the adjustment of the longitudinal profile. There is no way in which one could predict that the effect of a change in the independent controls would be better absorbed by a change in slope rather than by a change in the form of the cross section.

It may be, therefore, that a stream or reach may be virtually always adjusted, in the sense of being graded or in a steady state, without necessarily presenting the smooth longitudinal profile considered by the advocates of the geomorphic cycle as the hallmark of the "mature graded condition." The state of grade is thus analogous to the tendency for steady-state adjustment, it is perhaps always present and, therefore, this presence cannot be employed necessarily as an historical, or stage, characteristic. It is interesting that the concept of the vegetational "climax," which has often been compared to that of grade, has passed through a somewhat similar metamorphosis. The original idea of a progressive approach to a static equilibrium of the ecological assemblage has been challenged by the open system interpretation of Whittaker, with an historical link being provided by the "individualistic concept" of Gleason, much in the same way as Mackin's concept of grade links those of E'avis and Wolman.

The forms developed, together with the mutual adjustment of internal form elements and of related systems, are dependent on the flow of material and energy in the steady state. The laws of morphometry express one aspect of this relationship in geomorphology. In addition, adjustment of form elements implies a law of optimum size of a system and of elements within a system. This is mirrored by Gilbert's symmetrical migration of divides and by Schumm's constant of channel maintenance and is illustrated by Schumm's contrast between basin areas of differing order.

Although a steady state is in many respects a lime independent condition, it differs from the equilibrium of closed systems. A steady state means that the aspects of form are not static and unchanging, but that they are maintained in the flow of matter and energy traversing the system. An open system will, certain conditions presupposed, develop toward a steady state and therefore undergo changes in this process. Such changes imply changes in energy conditions and, connected with these, changes in the structures during the process. The trend toward, and the development of, a steady state demands not an equation of force: and resistance over the landscape, but that the forms within the landscape are so regulated that the resistance presented by the surface at any point is proportionate to the stress applied to it.

Erosion on a slope of homogeneous material with uniform vegetative cover will be most rapid where the erosional power of the runoff is greatest. This non-uniform erosional process will in time result in a more stable slope profile which would offer a uniform resistance to erosion.

In this way the transport of mass and energy (i.e., water and debris) is carried on in the most economical manner. With time, landscape mass is therefore being removed and progressive changes in at least some of the absolute geometrical properties of landscape, particularly relief, are inevitable. It is wrong, however, to assume, as Davis did, that all these properties are involved necessarily in this progressive, sequential change. To return briefly to the analogy of the bowl. If the rush of water through the outflow is capable of progressively enlarging the orifice, the increasing discharge at the outflow, uncompensated at the inflow, will cause the head of water in the bowl to decrease. This loss of head will itself, however, constantly tend to compensate the increasing outflow, but, if the enlargement of the outflow orifice proceeds, this is a losing battle and an important feature of the system will be the progressive and sequential loss of head. However, not all features of this system will reflect this progressive change of head, and, for example, the structure of the flow within the bowl will remain much the same while any head of water at all remains there. The dimensionless ratios between landscape forms, similarly, seem to express the steady state condition of adjusted forms from which mass is constantly being removed. The geometrical ratios which form the basis of the laws of morphometry, and the height-area ratios involved in the dimensionless, equilibrium hypsometric integral are examples of this adjustment.

In late mature and old stages of topography, despite the attainment of low relief, the hypsometirc curve shows no significant variations from the mature form, and a low integral results only where monadnocks remain. After monadnock masses are removed, the hypsometric curve may be expected to revert to a middle position with integrals in the general range of 40 to 60 percent.

In a drainage basin composed of homogenous material, in which no monadnocks would tend to form, it seems possible, therefore, that the dimensionless percentage volume of unconsumed mass (represented by the hypsometric integral) may achieve a time-independent value. It has been suggested, however, that the construction of the hypsometric curve may be so inherently restricted as to make the hypsometric integral insensitive to variations of an order which would be necessary to recognize such an equilibrium state. This steady state principle has been tentatively extended by Schumm to certain other aspects of drainage basin form.

The form of the typical basin at Perth Amboy changes most rapidly in the earliest stage of development. Relief and stream gradient increase rapidly to a point at which about 25 percent of the mass of the basin has been removed, then remains essentially constant. Because relief ratio [the ratio between total relief of a basin and the longest dimension of the basin parallel to the principal drainage] elsewhere has shown a close positive correlation with stream gradient, drainage density, and ground-slope angles, stage of development might be expected to have little effect on any of these values once the relief ratio has become constant.

In the steady state of landscape development, therefore, force and resistance are not equated (which would imply no absolute form change), but balanced in an aerial sense, such that force may still exceed resistance and cause mass to be removed. Now, as has been pointed out, removal of mass under steady-state conditions must imply some progressive changes in certain absolute geometrical properties of a landscape, notably a decrease in average relief, but by no means all such properties need respond in this simple manner to the progressive removal of mass. The existence, for example, of the optimum magnitude principle for individual systems, or subsystems, implies that if the available energy within the system is sufficient to impose the optimum magnitude on that system, this magnitude will be maintained throughout a period of time and will not always be susceptible to a progressive, sequential change. Thus, Strahler has indicated that erosional slopes which are being forced to their maximum angle of repose by aggressive basal stream action will, of necessity, retain this maximum angle despite the progressive removal of mass with time.

Total energy is made up of interchangeable potential energy and flux, or kinetic, energy and even if the potential energy component decreases within an open system due to its general reduction, in other words along with a continual change in one aspect of form (i.e., relief), the residual flux energy may be of such overriding importance as to effectively maintain a steady state of operation. In practice the steady state is seldom, if ever, characterized by exact equilibrium, but simply by a tendency to attain it. This is partly due to the constant energy changes which are themselves characteristic of many open system operations, but the steady state condition of tendency toward attainment of equilibrium is a necessary prerequisite, according to Von Bertalanffy for the system to perform work at all. Now, once a steady state has been established, the influence of the initial system conditions vanishes and, with it, the evidence for a previous history of the system (i.e., was our bowl full or empty at the start?). Indeed, in terms of analyzing the causes of phenomena which exhibit a marked steady state tendency, considerations regarding previous history become not only hypothetical, but largely irrelevant. This concept contrasts strikingly with the historical view of development which is fostered by closed system thinking. Wooldridge and Linton have gone so far as to say that.

Any such close comprehension of the terrain can be obtained in one way only, by tracing its evolution. An even more extreme statement of the same philosophy has been made by Wooldridge and Goldring. The physical landscape, including the vegetation cover, is the record of processes and the whole of the evidence for its evolution is contained in the landscape itself.

The whole matter hinges on the rapidity with which landscape features become adjusted to energy flow, which may itself be susceptible to rapid changes, particularly during the rather abnormal latest geologic period of earth history. Obviously, most existing features are the product of both past and reasonably contemporary energy conditions, and the degree to which these latter conditions have gained ascendancy over the former is largely a function of the ratio between the amount of present energy application and the strength (whatever this may mean) of the landscape materials. Thus, the geometry of stream channels and the morphometry of weak clay bad lands show remarkable adjustments to contemporary processes on whatever time level the action of these processes may be denned whereas, at the other end of the energy/resistance scale, erosion surfaces cut in resistant rock and exposed to the low present energy levels associated with the erosional processes of certain areas of tropical Africa can only be understood on the basis of past conditions. Between these two extremes lies the major part of the subject matter of geomorphology including considerations of slope development, and it is here where the apparent dichotomy between the two systematic approaches to the same phenomena, termed by Bucher "timebound" and "timeless," is most acute. In a related context, the problem of timebound-versus-timeless phenomena becomes especially obvious when rates of change and the ability to adjust are underestimated, as when vegetational assemblages have been correlated with the assumed stages of geomorphic history in the folded Appalachians by Braun and in Brazil by Cole.

One can appreciate that in areas where good evidence for a previous landscape history still remains the historical approach may be extremely productive, as exemplified by the work of Wooldridge and Linton on south-eastern England. However, in many (if not most) areas the condition is one of massive removal of past evidence and of tendency toward adjustment with progressively contemporaneous conditions. It is an impossibly restricted view, therefore, to imagine a universal approach to landform study being based only upon considerations of historical development.

Another characteristic of the open system is that negative entropy, or free energy, can be imported into it because of its very nature. Therefore, the open system is not defined by the trend toward maximum entropy. Open systems thus may maintain their organization and regularity of form, in a continual exchange of their component materials. They may even develop toward higher order, heterogeneity, hierarchical differentiation and organization. This is mirrored in geomorphology by the characteristic development of interrelated drainage forms, and goes along with a concept of progressive segregation. This, to a minor extent, insofar, as, with time, rates of interactions between form elements in an open system may tend to decrease. Therefore, it is quite reasonable to assume that mutual adjustments of form within geomorphic systems might be more difficult of accomplishment and delayed where the relief, through its influence over the potential energy of the system, is low rather than where there is a higher potential energy in the system.

Steady-state conditions can be interrupted by a disturbance in the energy flow or in the resistance, leading to form adjustments allowing a new steady state to be approached. These adjustments, however, do imply a consumption of energy and there is a "cost of transition" from one steady state to another. A particular geomorphic instance of this dissipation might be presented by the phenomenon of "overshooting" where active, but sporadic, processes are operating on weak materials, as instanced when the failure of steep slopes reduces them to inclinations very much below their repose angles, and by the excessive cutting and subsequent filling of alluvial channels associated with flash floods.

The dynamic equilibrium of the steady state manifests itself in a tendency toward a mean condition, of unit forms, recognizable statistically, about which variations may take place over periods of time with fluctuations in the energy flow. These periods of time may in some instances be of very short duration, and the fluctuations of transverse stream profiles are measurable in the days, or even minutes, during which changes of discharge occur. These constant adjustments to new steady-state conditions may be superimposed on a general tendency for change possibly associated with the reduction of average relief through time. This general relief change, however, does not imply a sympathetic change of all the other features of landscape geometry. As has been demonstrated by Strahler and Melton for example, drainage density is controlled by a number of factors of which relief is only one. Recent work seems to be indicating that relief (naturally including considerations of average land slope) probably has only a relatively small influence over drainage density, which may be masked or negated altogether by the other more important factors (for example, rainfall intensity and surface resistance) which are not so obviously susceptible to changes with time. Denbigh, Hicks and Page have pointed out that: Quite large changes of environment may take place, without the need for more than a small internal readjustment.

Horton did not believe, as did Glock that drainage density could be employed as a measure of landscape "age," and, indeed, it is not difficult to entertain the possibility that certain features of landscape geometry may be relatively unchanging, in actual dimensional magnitude as well as in dimensionless ratio, throughout long periods of erosional history.

For many landscape units, changes on either level are slow, or in some instances non-existent. Under steady state conditions, therefore, corresponding local morphometric units will, as regards their form and magnitude, tend to crowd around a very significant mean value, imparting to a geomorphic region its aspects of uniformity. Strahler's "law of constancy of slopes" is an expression of one phase of this adjustment. It is interesting that the general principle of the operation of a steady state condition was intuitively recognized long ago by Play fair.

The geological system of Dr.Button, resembles, in many respects, that which appears to preside over the heavenly motions. In both, we perceive continual vicissitude and change, but confined within certain limits, and never departing far from a certain mean condition, which is such, that in the lapse of time, the deviations from it on one side, must become just equal to the deviations from it on the other.

Often the achievement of exact equilibrium in nature occurs only momentarily as variations about the mean take place and in these instances the existence of the steady state can only be recognized statistically. In the study of landscape, the steady state condition indicated by discrete, close and recognizable statistical groupings of similar units is characteristic of regions of uniform ratios between process and surface resistance.

Davis' view of landscape evolution was that the passage of time, of necessity, imprinted recognizable, significant and progressive changes, on every facet of landscape geometry. The recognition, however, that landscape forms represent a steady-state adjustment with respect to a multiplicity of controlling factors obliges one to take a less rigid view of the evolutionary aspects of geomorphology. When a geometrical form is controlled by a number of factors, any change of form with the passage of time is entirely dependent upon the net result of the effect of time upon those factors. Some factors are profoundly affected by the passage of time, others are not; some factors act directly (using the term in the mathematical sense) upon the form, others inversely some factors exercise an important control over form aspects, others a less important one. Thus, if a particular geometrical feature of landscape is primarily controlled by a factor the action of which does not change greatly with time, or if the changes of factors having direct and inverse controls tend to cancel out the net effect of the changes, then the resulting variation in geometry may itself -be small perhaps insignificant.

A last important characteristic of open systems is that they are capable of behaving "equifinally" in other words different initial conditions can lead to similar end results. Davisian (closed system) thinking is instinctively opposed to this view, and the immediate and facile assumption, for example, that most breaks of stream slope are only referable to a polycyclic mechanism is an illustration of the one cause-one effect mentality. The concept of equifinality accentuates the multivariate nature of most geomorphic processes and militates against the unidirectional inevitability of the closed system cyclic approach of Davis. The approach contrasts strikingly with that of Gilbert.

Phenomena are arranged in chains of necessary sequence. In such a chain each link is the necessary consequent of that which precedes and the necessary antecedent of that which follows. If we examine any link of the chain, we find it has more than one antecedent and more than one consequent. Antecedent and consequent relations are therefore not merely linear, but constitute a plexus; and this plexus pervades nature.

To sum up, the real value of the open system approach to geomorphology is: Firstly, that it throws the emphasis on the recognition of the adjustment, or the universal tendency toward adjustment, between form and process. Both form and process are studied, therefore, in equal measure, so avoiding the pitfall of Davis and his more recent associates of the complete ignoring of process in geomorphology.

In a graded drainage system the steady state manifests itself in the development of certain topographic form characteristics which achieve a time-independent condition. Erosional and transportational processes meanwhile produce a steady flow (averaged over a period of years or tens of years) of water and waste from and through the landform system. Over the long span of the erosion cycle continual adjustment of the components in the steady state is required as relief lowers and available energy diminishes. The forms will likewise show a slow evolution. Applied to erosion processes and forms, the concept of the steady state in an open system focuses attention upon the relationship between dynamics and morphology.

The relation between process and form lies close to the heart of geomorphology and, in practice, the two are often so intimately linked that the problem of cause and effect may present the features of the "hen and the egg." Approach from either direction is valuable, however, for knowledge of form aids in the understanding of process, and studies of process help in the clearer perception of the significant aspects of form.

The study of form may be descriptive merely, or it may become analytical. We begin by describing the shape of an object in the simple words of common speech. We end by defining it in the precise language of mathematics; and the one method tends to follow the other in strict scientific order and historical continuity. The mathematical definition of a "form" has a quality of precision which was quite lacking in our earlier stage of mere description (employing means which) are so pregnant with meaning that thought itself is economized.

We are apt to think of mathematical definitions as too strict and rigid for common use, but their rigour is combined with all but endless freedom we reach through mathematical analysis to mathematical synthesis. We discover homologies or identities which were not obvious before, and which our description obscured rather than revealed.

Once more, and this is the greatest gain of all, we pass quickly and easily from the mathematical concept of form in its statical aspect to form in its dynamical relations: we rise from the conception of form to an understanding of the forces which gave rise to it; and in the representation of form and in the comparison of kindred forms, we see in the one case a diagram of forces in equilibrium, and in the other case we discern the magnitude and the direction of the forces which have sufficed to convert the one form into the other.

Every natural phenomenon, however simple, is really composite, and every visible action and effect is a summation of countless subordinate actions. Here mathematics shows her peculiar power, to combine and generalize.

A large part of the neglect and suspicion of mathematical methods in morphology is due to an ingrained and deep-seated belief that even when we seem to discern a regular mathematical figure in an organism (the form) which we so recognise merely resembles, but is never entirely explained by, its mathematical analogue; in short, that the details in which the figure differs from its mathematical prototype are more important and more interesting than the features in which it agrees; and even that the peculiar aesthetic pleasure with which we regard a living thing is somehow bound up with the departure from mathematical regularity which it manifests as a peculiar attribute of life. We may be dismayed too easily my contingencies which are nothing short of irrelevant compared to the main issue; there is a principle of negligibility.

If no chain hangs in a perfect catenary and no raindrop is a perfect sphere, this is for the reason that forces and resistances other than the main one are inevitably at work but it is for the mathematician to unravel the conflicting forces which are at work together. And this process of investigation may lead us on step by step to new phenomena, as it has done in physics, where sometimes a knowledge of form leads us to the interpretation of forces, and at other times a knowledge of the forces at work guides us towards a better insight into form.

Secondly, open-system thinking directs the investigation toward the essentially multivariate character of geomorphic phenomena. It is of interest to note that the physical, and the resulting psychological, inability of geographers to handle successfully the simultaneous operation of a number of causes contributing to a given effect has been one of the greatest impediments to the advancement of their discipline. This inability has prompted, at worst a uncaused determinism and, at best, an unrealistic concentration upon one or two contributing factors at the expense of others. Davis' preoccupation with "stage" in geomorphology has been paralleled, for example, by an undue emphasis on the part of some economic geographers upon the factor of "distance" in many analyses of economic location.

Thirdly, it allows a more liberal view of changes of form with time, so as to include the possibility of non-significant or non-progressive changes of certain aspects of landscape form through time.

Fourthly, while not denying the value of the historical approach to landform development in those areas to which the application of this framework of study is appropriate, open-system thinking fosters, a less rigid view regarding the aims and methods of geomorphology than that which appears to be held by proponents of the historical approach. It embraces naturally within its general framework the forms possessing relict facets, those indeed which form the basis for the present studies of denudation chronology, under the general category of the "inequilibrium" forms of Strahler. There is no uniquely correct method of treatment for a given body of information, and Postan has been at pains to demonstrate the purely subjective distinction which exists between alternative explanations of phenomena on an immediately causal or generic basis, as against an historical or biographical one.

For the frontier they draw separates not the different compartments of the universe but merely the different mental attitudes to the universe as a whole. What makes the material fact a fit object for scientific study is that men are prepared to treat it as an instance of a generic series. What makes a social phenomenon an historical event is that men ask about it individual or, so to speak, biographical questions. But there is no reason why the process should not be reversed; why we should not ask generic questions about historical events or should not write individual biographies of physical objects. Here Spinoza's argument still holds. The fall of a brick can be treated as a mere instance of the general study of falling bricks, in which case it is a material fact, and part and parcel of a scientific enquiry. But it is equally possible to conceive a special interest in a particular brick and ask why that individual brick behaved as it did at the unique moment of its fall. And the brick will then become an historical event. Newton must have been confronted with something of the same choice on the famous day when he sat under the fabulous apple tree. Had he asked himself the obvious question, why did that particular apple choose that unrepeatable instant to fall on that unique head, he might have written the history of an apple. Instead of which he asked himself why apples fell and produced the theory of gravitation. The decision was not the apple's but Newton's. Davis was metaphorically struck by landscape and chose to write a history of it.

Fifthly, the open-system mentality directs the study of geomorphology to the whole landscape assemblage, rather than simply to the often minute elements of landscape having supposed evolutionary significance.

Sixthly, the open-system approach encourages rigorous geomorphic studies to be carried out in those regions and perhaps these are in the majority where the evidence for a previous protracted erosional history is blurred, or has been removed altogether.

Lastly, open-system thinking, when applied to geomorphology, has application within the general framework of geography; for geomorphology has always influenced geographical thinking to a great, and possibly excessive, degree. Open system thinking is characteristically less rigidly deterministic in a causative and time sense than the closed-system approach. The application of this closed-system approach to problems of human geography is extremely dangerous because, of its nature, it directs the emphasis toward a narrow determinism, and encourages a concentration upon closed boundary conditions, upon the tendency toward homogeneity and upon the levelling down of differences. Open-system thinking, however, directs attention to the heterogeneity of spatial organization, to the creation of segregation, and to the increasingly hierarchical differentiation which often takes place with time.

Hardware Models in Geomorphology

Hardware modelling in geomorphology may be defined as the study, under monitored or controlled conditions, of a physical representation of a selected geomorphic feature. There are three broad classes of hardware model: segments of unscaled reality (e.g., a block of soil brought into the laboratory, or a `study area'), scale models, and analogue models (in which aspects of the form and/or processes of a natural geomorphic phenomenon are reproduced by forces, materials, and processes quite dissimilar to those in nature - as when a thread of water winding down the underside of a glass plate is used to demonstrate river meanders. The main drawback with hardware models is that they cannot be used on their own to develop theories or laws, but they have many applications in testing and verifying these once formulated.

Scaled Physical Models

Physical models are scaled representations of a full-scale physical system which can be applied to inform our understanding of geomorphic process-form interactions. Physical and experimental modelling has been used extensively and has been proven to be of critical importance to the geomorphological user.

Model Typology

The choice of physical model type is dependent on various factors including the project objectives and rationale, as well as cost and space limitations. Additionally, the purpose of the physical model will control the type of model that is used, with models generally being constructed for:

- Research purposes.
- Communication and education purposes.
- Informing decisions or providing foresight.

Two types of boundary condition have been recognised within physical models: fixed-bed, where the model boundaries are non-erodible and no sediment transport can occur; and, moveable-bed, where substrate is free to move within a constrained or non-constrained channel Physical models can be loosely divided into a number of categories, including:

- Scaled models,

- Froude number scaled models,
- Distorted scale models,
- Analogue models,
- 1:1 replica models.

Despite this classification scheme, some overlap may exist, for example, an analogue model may exhibit characteristics associated with all of the other categories.

Table: Purposes or aims of physical models.

Project aims	Sub-discipline of geomorphology	
Research tools to study process-form interactions	Fluvial	Influence of in channel/floodplain vegetation on river morphology.
	Fluvial	Investigations into alluvial fan dynamics and evolution.
	Hillslope	Investigations into hillslope-channel coupling processes.
	Glacial	Investigations into jökulhaups with different hydrograph shapes and their subsequent impacts.
	Aeolian/dryland	Wind-tunnel tests on aeolian transport of different sized sand grains under varying wind velocities.
Education, demonstration and communication tools	Fluvial	Micro-model flume to communicate channel avulsion processes to the public, students or stakeholders.
	Glacial	Glacier dynamics under changing climate experiments using PVC piping valley and viscous flow medium
Screening tools to seek alternative approaches/improve understanding	Fluvial	Use of physical models to inform understanding of the downstream and upstream impacts of channel impoundment or dam removal.
	Fluvial	Process understanding of ice jams at river confluences.

Scaled Models

Scaled physical models that are built and function at reduced scale (or enlarged scale in some cases) are an important type of physical model for examining and measuring processes which are difficult to observe in reality. Scaled models allow geomorphological users to overcome the inherent obstacles associated with investigating physical systems such as the long spatiotemporal timescales involved and problems associated with working in a naturally variable environment. Scaled physical models conform to scale ratios, shown by equation:

$$N_x = \frac{x_p}{x_m} = \frac{\text{Value of X in Prototype}}{\text{Value of X in Model}}$$

Where N_x is the actual-to-model scale ratio of parameter x (which may represent width, depth, length, grain size, time, diameter etc.), and where p and m represent the actual/original system and model, respectively. Thus, if a river reach has a length of 300 m in reality but this is scaled to 3 m under modelled conditions, the model length is said to be scaled by 1:100.

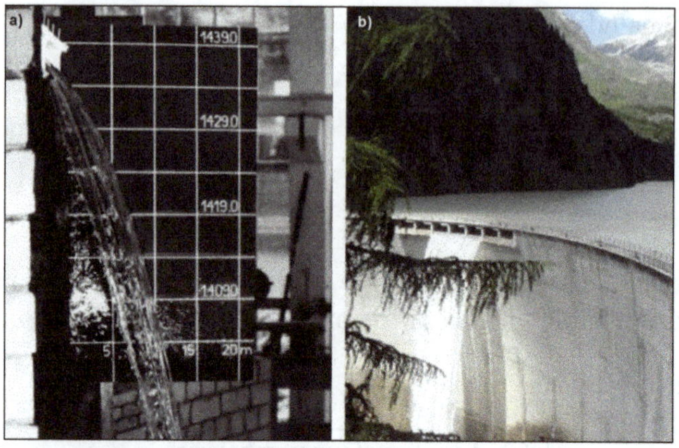

Overflow spillway of Gibidum Dam, Switzerland. (a) 1:30 scaled physical hydraulic model. (b) real-world, full scale photograph of Gibidum Dam, which the model is based on.

Scaling occurs in all physical models to varying extents however, modellers must be cautious when down scaling a model too much from the real world system. Maynord evaluated a 'micro-model' river system (1:14,000 horizontal, 1:1,200 vertical scale) with a river channel width as small as 4 cm. It was demonstrated that using large scaling factors resulted in models becoming incomparable to the hydrodynamic processes occurring in reality. For example, a river channel width of 560 m in reality cannot be represented as 4 cm width in a physical model due to the significantly different hydrodynamic processes occurring. Additionally, when scaling particle sizes, users must be cautious of cohesive forces becoming a dominant factor in the model while being absent or negligible in reality. This may result in 'micro-model' systems losing their predictive capabilities and becoming qualitative rather than quantitative. Whether this is disadvantageous or not depends on the researcher's aims; qualitative modelling may still be useful for demonstration, education and communication purposes, as well as a rapid, and visual screening tool to inform research direction.

Fluvial geomorphology in a box. Micro-model river system allowing the study of fluvial dynamics, similar to the one used by Maynord.

Froude Number Scaled Models

A true scaled model requires perfect geometric, kinematic and dynamic similitude, something that cannot be achieved when using the same fluid as in the real world system due to equivalent gravitational and fluid motion forces. Therefore, one or more variables must be relaxed in order to achieve model-field similitude. Froude number scaling can be applied, whereby the Reynolds number (Re), a dimensionless quantity used to quantify turbulence rate, is relaxed while correctly scaling the Froude number (Fr), a measurement of different flow states, e.g. subcritical, critical and supercritical. If this was not done, experimental models involving water would have a significantly lower Reynolds number than their counterpart fullscale system, resulting in a lack of similarity between model and reality. Instead of having to reduce the viscosity of fluid or to build a 1:1 replica model, Froude number scaling allows a smaller-scale physical model involving fluid flow to produce similar characteristics to its real-world counterpart. For free surface flow, gravitational forces are dominant. Therefore, hydraulic similarity can be established by equating the ratio of gravitational forces to that of inertial forces. Examples of the effectiveness of Froude number scaling include the work of Ashmore who classified the mechanisms of river braiding and controls on bar formation and related the internal generation of bedload pulses to channel avulsion. A Froude number scaled model was applied to produce compariable results within the physical model to that of the field counterpart. In Ashmore's study, the use of a Froude number scaled model allowed an understanding of braided channel morphology, flow characteristics and bedload movement where field measurements and observations were challenging due to the large spatio-temporal scales involved.

Distorted Scale Models

Scaled physical models adhere to dimensional scaling of all axes to the same ratio, whereby all attributes within the model are geometrically similar to the original system. However, it is also common for scaled physical models to be geometrically distorted and skewed. Geometrically distorted scaled models, where the scaling of a model's vertical to horizontal scaling ratio differs, are especially important to the geomorphological user when large spatial scales that cannot correctly be replicated under laboratory conditions or fine sediment sizes are involved. Distorted scale models enable small physical models to be built or large physical systems to be modelled. Additionally, distorted scale models may be applied to avoid problem of water or fluids behaving viscid at rigid boundaries. Distorted scale model experiments may involve variables such as width, length, slope and/or grain size/density adhering to differing scaling factors. For example, McCollum used a distorted flume to understand sediment transport dynamics along a 7km river reach which experienced significant rates of sedimentation. Because of the impracticability of reproducing a 7 km flume under laboratory conditions and because conducting field studies would not allow experimental control over system variables (e.g. slope and/or discharge), a distorted flume with horizontal and vertical scaling ratios of 1:120 and 1:80 respectively was used. Furthermore, crushed coal was used to avoid unrealistic cohesion within the scaled model to ensure the distorted scale model produced a similar response to the field system. The San Francisco Bay Model, a working hydraulic model of the San Francisco Bay and Sacramento-San Joaquin River Delta system is also an example of a geometrically skewed physical model, with horizontal and vertical scaling being 1:1,000 and 1:100, respectively. Furthermore, the model operates at a temporal scale of 1:100, with one diurnal cycle being represented in approximately 15 minutes.

Analogue 'Similarity of Process' Models

Analogue models are models that reproduce certain features of a natural system even though the processes, forms, dynamics, behaviour, materials and/or geometries do not conform to scaling ratios of the actual system. These are useful when true similarity between model and original system is unachievable or unnecessary. Analogue models may appear to be considerably different from the original field system but are based upon Hooke's similarity of process' concept, whereby the laboratory setup is considered a small system in its own right, rather than a scaled down reality. Seen as models and not miniature reproductions, analogue models should be treated as real, albeit simple physical systems relying on the premise that processes occurring within a natural system will be comparable to those within a laboratory environment. This allows analogue models to output detailed and transferable process understanding rather than an understanding that is case study specific. Advantages of using analogue models over other physical model types include their potentially rapid setup times, their ability to conduct prompt scenario testing and the reduced quantitative extrapolation required to make conclusions upon. However, analogue models may encounter difficulties with relating measurements and results obtained within the modelling environment to real world situations.

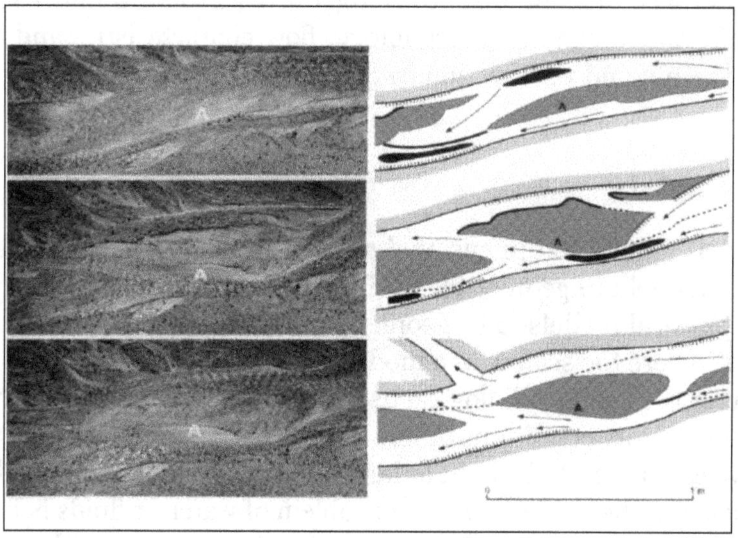

Figure shows flume study of a braided river system showing medial bar destruction caused by longitudinal translation and change in total discharge of an upstream confluence, demonstrating the influence that physical modelling has had upon braided river system understanding. Elapsed time of physical model simulation is 1 hour.

Replica Models

Some systems are small enough to be replicated in a laboratory and can be simulated at 1:1 scale (i.e. maintaining the exact dimensions of the studied physical system). This allows the studied system to be modelled under laboratory conditions with little or no difference. This has a number of advantages, such as a large degree of experimental control over model parameters. However, 1:1 replica models are not suitable for large-scale geomorphological systems due to space limitations within a laboratory setting.

Numerous 1:1 flume studies exist, such as researchers who used observations from an unscaled flume to understand fluvial bedload abrasion rates. Additionally, 1:1 experiments can be conducted in the field under natural settings but with controlled inputs and conditions. The Outdoor Stream Laboratory, Minnesota, is a field-size reproduction of a fluvial system which is part of the Saint Anthony Falls Laboratory. This allows an understanding of the underlying physical, biological and chemical mechanisms that govern stream and riparian processes and their response to natural and human disturbances under controlled conditions, e.g. steady and unsteady inlet hydrographs, to simulate overbank flood dynamics. Additionally, the Laboratory for Experimental Geomorphology in Leuven, as well as Moss and Walker's experiments, have conducted 1:1 laboratory experiments focusing on a number of in-situ surface erosion processes and their relationship to surface material properties. These include soil and rain splash erosion plot studies and tillage experiments and rill development.

Principles of Physical Modelling

Despite the long history of physical modelling studies (Da Vinci used physical models to observe flow characteristics in the 1500s, Reynolds conducted moveable bed models of the River Mersey, UK in 1885 and the US Army Corps of Engineers commissioned multiple large-scale physical modelling experiments from the 1920s onwards, e.g. Coastal Engineering Research Centre and the Waterways Experiment Station; Markle, 1989) there is currently no established framework for conducting experimental and physical modelling studies. At present, laboratories using physical models adopt their own individual approaches based on institutional experience or communication with other similar projects. Because physical models may be used for a variety of geomorphological applications, procedures vary significantly between projects. Despite this, there are a number of unifying principles that all physical modelling projects should consider, including:

- Similitude requirements,
- Dimensional analysis,
- The materials that are used.

Similitude

Similitude, also known as similarity, involves the model resembling and being correspondent to the system which the model is based upon. Similitude can be divided into three types: geometric (form); kinematic (motion) and dynamic (force) similitude. Firstly, geometric similitude involves the physical model being similar to its real world counterpart in regards to its dimensions and measurements, involving similarity in form. Therefore, a reduced or enlarged reproduction of the studied physical system is needed to achieve geometric similitude. Secondly, kinematic similitude involves similarity in motion being achieved between the model and real world system, with the ratio of movement in both systems being directly proportional. As a result, true kinematic similitude produces model particle/flow pathways that are geometrically similar to the actual physical system. Finally, dynamic similitude involves the proportion of relevant forces acting upon fluid flows and boundary surfaces being comparable between the model and full scale systems. This produces length, mass and time measurements that are proportionate, implying a constant ratio of forces between both systems. To achieve dynamic similitude, both geometric and kinematic similitude is required.

The degree to which similitude is satisfied is dependent upon:

- The aims and objectives of the researcher.
- Whether the physical model is generating qualitative or quantitative data.
- Whether the model can be calibrated and adjusted using existing data or models.

Scale effects which arise due to force ratios being incomparable between a model and its real-world counterpart, and laboratory effects which arise due to the inability of a laboratory to simulate the correct forcing conditions and model boundaries may hinder dynamic similitude. Full dynamic similitude within a scaled physical model is often difficult, if not impossible, to achieve due to force vectors being required to be equal between both systems. Scaled models involving water are unable to achieve dynamic similitude as the model fluid would be needed to have a different viscosity to its real-world counterpart. To avoid using a different liquid from the real-world system, users may either use a physical model that functions at full scale (1:1), or use Froude number scaled models to relax the similitude requirements. Because all geomorphic physical models are unique, achieving similarity between model and actual system can be relaxed as long as the similitude requirements are justified and reasonable.

Dimensional Analysis

For scaled physical models to be representative of their full-scale system, quantities measured may adhere to scaling laws. Neglect of scaling considerations may render model results meaningless for scientific interpretation or prevent the model from correctly predicting process-form interactions at the actual system scale. Scale models are based on similitude theory (above). One method of achieving similitude is by producing a series of dimensionless parameters that are able to form relationships between physical processes. Yalin notes that the dimension of any physical system can be characterised in terms of its fundamental dimensions; length, time and mass. Using this concept, dimensional analysis, involving the examination of the relationships between different physical parameters by identifying their fundamental dimensions (time, mass, length) to determine the derived quantities (e.g. area, volume, force, velocity, frequency which is a function of the fundamental dimensions) can be applied.

Table: Dimensions of physical entities using a mass system of units.

Physical property	Dimensions	Type of quantity
Fundamental quantities		
Time	[T]	-
Mass	[M]	-
Length	[L]	Geometric
Temperature	[ø]	-
Angle	[1]	(Supplementary)
Derived quantities		
Area	[L^2]	Geometric
Volume	[L^3]	Geometric
Force	[MLT^{-2}]	Dynamic

Velocity	$[LT^{-1}]$	Kinematic
Acceleration	$[LT^{-2}]$	Kinematic
Volumetric Flow Rate	$[L^3 T^{-1}]$	Kinematic
Strain	[1]	Dimensionless

Materials in Geomorphological Research

The use of materials in geomorphological research is often highly project-specific. Because of this, a few case studies and the author's experience have been highlighted allowing users to make informed but not constrained decisions.

Physical models may use the exact materials as their real world counterpart, e.g. soil, gravel, flora and fauna in nature and in the physical model. Conversely, physical models may use surrogate/proxy materials that differ from the actual materials present within the physical system. This includes the scaling down of materials within a physical modelling environment and using alternative materials which mimic or substitute the use of the actual materials present in nature. Examples include:

- Using sand instead of gravel.
- Substituting live vegetation with a smaller species or using an artificial surrogate.
- Using sponge to represent soil.
- Using lighter bed materials like pumice or charcoal to represent larger clasts.
- Using of fluids with differing viscosities to that of the real world system.

Using similar sediment in a scaled physical model to that found in an actual system may lead to responses in model behaviour that are not comparable to the real world counterpart. This has been documented in scaled flume studies, where using similar sediments resulted in the formation of ripples that had no equivalent in the field. This affected the flow dynamics, leading to supercritical flow, hydraulic jumps and standing waves that influenced bed morphology and rates of erosion. To avoid this, lighter bed material such as pumice, charcoal or sand can be used.

Vegetation is commonly used in physical models. When using vegetation in physical models either:

- Artificial/surrogate plants.
- Scaled, smaller species.
- Natural vegetation can be used.

When using artificially scaled substitutes, careful consideration must be taken to ensure that these are representative of the actual physical system. Using artificial vegetation has the benefit that it is inert, controllable and easy to use. However, the user must be aware of the limitations associated with using a proxy material to represent a highly variable component of a physical system. Limitations of using artificial vegetation include:

- Misrepresentation of a plant surrogate to replicate the behaviour of natural vegetation.
- The vegetation characteristics (e.g. Flexibility and density) not being comparable between model and nature.

Natural vegetation has the benefit that it is directly comparable to that of a natural system, however, it is also highly variable and potentially difficult to maintain in laboratory conditions. Scale is also important to consider. Modellers would need to substitute larger vegetation types, e.g. trees, with smaller saplings or shrubs due to the space limitations associated with using an experimental set-up. Therefore, the plant materials used depends upon number factors including the scale of the flume and the purpose of the experiments.

Table: Choice of plants in physical modelling.

	Choice of plant	Purpose
Artificial/ surrogate	Rods or wooden dowels.	Stem density effects on drag and flow resistance. Flow resistance on flood plains.
	Rods with strips, plastic strips or strips with foliage attached.	Flow structures, vegetation-flow interactions.
	Plastic bushes or grasses.	Floodplain roughness on flow structures, bed-forms and sediment transport rates.
Scaled	Smaller vegetation (e.g. Medicago sativa)	Flow resistance and controls on stream morphodynamics.
Natural	Natural, full-scale vegetation, such as grass, shrubs or trees.	Flow resistance, plant-flow interactions.

Cellulose sponge can be used as a proxy material for soil. Researchers investigated the validity of the simile 'soils are like sponges', demonstrating through experimental methods that sponges store and release water in much the same ways that soils do, with cellulose sponge having intermediate hydrological characteristics to peat and topsoil. Although it is identified that sponge has a higher water retention capacity than soil, over 2.5 times than peat soils, Researchers do not address the fact that sponge could be scaled volumetrically by thickness/depth to account for this additional storage. Scaling by storage capacity allows sponge to provide a clean and non-erodible medium to investigate runoff and infiltration processes. Sponge can also be compressed to remove stored water, allowing rapid repetition of experimental runs. This highlights the potential benefits of using sponge as a proxy material in physical models. Although sponge allows numerous benefits to the physical modeller, sponge would not be suitable at 1:1 scale; using soil would produce more realistic outputs and avoids using proxy/surrogate materials.

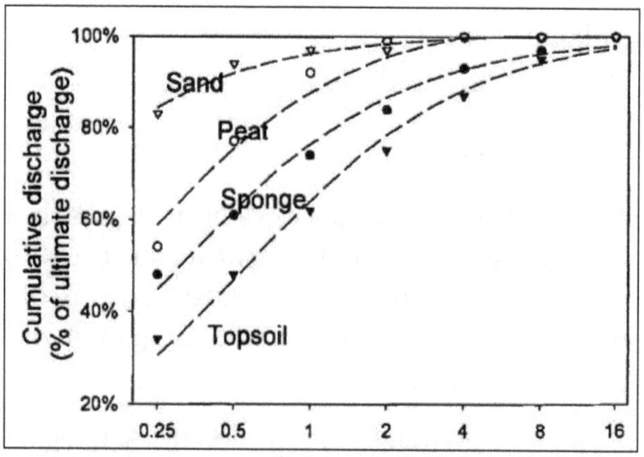

Figure shows results from laboratory testing of cumulative discharge of water from four media, demonstrating sponge's ability to act as a proxy material for soil. Other experimental tests including rates of flow, water potential curves and gravimetric water contents in cellulose sponge were also intermediate to that of topsoil and peat.

Applications of Physical Models

Physical models have an important role in geomorphological research. To demonstrate the scope and potential of physical modelling in geomorphological research, fluvial, glacial, aeolian and bio-geomorphological case studies have been highlighted. Physical models have also been used within coastal and hillslopegeomorphology/soil erosion studies.

Fluvial Geomorphology

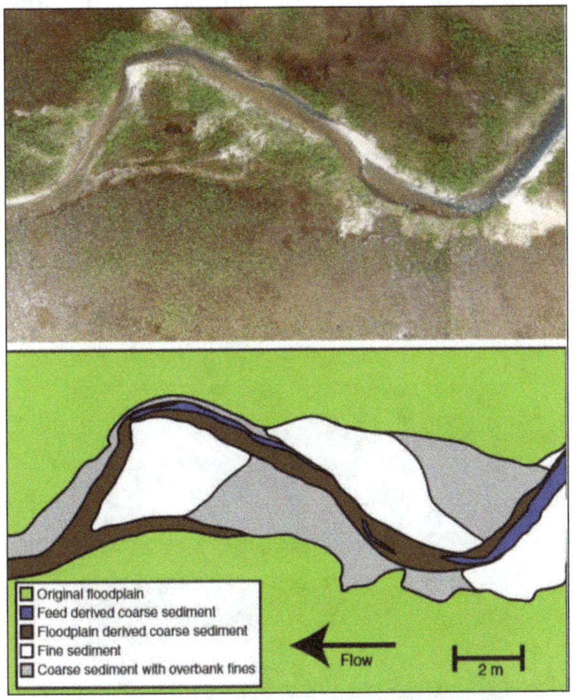

Figure shows sediment in second and third bars downstream from the flume inlet. Fine sediment is mapped where the majority of the floodplain thickness was fine sediment. Accumulation of organic matter from the dead alfalfa makes some of the bar appear brown where it is primarily fine sediment.

Physical models have been used extensively in fluvial geomorphology to understand a plethora of geomorphic processes, including sediment transport, river channel change and the influence of vegetation on channel adjustment. Fluvial geomorphological research using experimental methods is predominantly flume-based. Fluvial physical models were of crucial importance in the work of Hooke, who used laboratory streams to develop the 'similarity of process' model concept, as well as a number of US Army Corps of Engineers projects, e.g. the SEDflume project, a 6m long mobile flume which can analyse fluvial sediment sorting, and the Ice Harbour Lock and Dam Physical Model Study, a 1:55 scale dam commissioned to understand the downstream impacts of river impoundment. Researchers applied an experimental basin model of an aggrading braided river channel to investigate the relationship between the frequency of channel avulsion, the duration of time that the braid plain is occupied by flow, the spatial pattern of sedimentation and how these respond to a change in sediment supply. Results obtained from the physical model demonstrated a strong positive relationship between sediment supply and channel avulsion rates. Results attained within the physical modelling environment were also able to be extrapolated to real world examples to gain an understanding of braided river sedimentation.

Glacial Geomorphology

Published research on physical models in glacial geomorphology is sparse. The few studies that exist include Rushmer, who applied experimental flume methods to study the impact of glacial outburst floods with differing hydrographs and researchers who used physical modelling to investigate the influence of bedrock topography and ablation on ice flow direction and velocity using silicone gel. This study confirmed current conceptual models of ice flow around obstacles, demonstrating that variations in bed topography and internal layers of the ice are strongly influenced by the presence and height of bedrock obstacles.

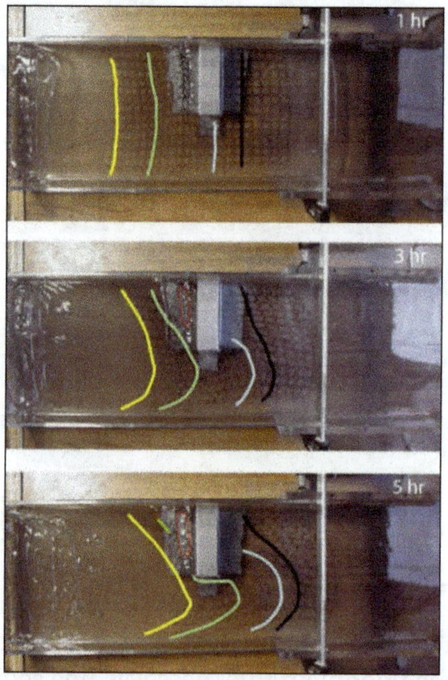

Physical model of a glacier, showing progressive deformation of silicone gel, an ice surrogate material, around an obstacle.

Glacial geomorphology is generally investigated using numerical, rather than physical models to describe relationships between mass balance, ice dynamics and climate, however, some aspects of glacial behaviour can be simulated using physical models, such as controls on ice melting, ice flow and sub-glacial erosion and sediment transport processes. Despite this lack of research, researchers express that physical models exhibit numerous opportunities for the glacial geomorphologist, such as the ability to isolate variables and study long spatio-temporal scales.

Aeolian Geomorphology

Theoretical understanding and the development of numerical models of Aeolian processes often contain empirical coefficients that need to be determined using wind tunnel tests, where variables such as grain size and wind speed/direction can be systematically controlled to investigate interactions. Authors have applied experimental wind tunnel tests to understand the relationships between flow velocities and sediment entrainment under differing wind velocity, grain size and moisture scenarios. These studies have confirmed the importance of using physical models to understand aeolian mechanisms.

Biogeomorphology

Biogeomorphology, the study of the interactions between flora and fauna and the development of landforms, is an emerging topic within geomorphology. Using flume experiments, researchers conducted ecological experiments to demonstrate that crayfish activity significantly affects sand and gravel erosion by increasing bed roughness, decreasing bedform height and altering the pool-riffle sequence downstream. More recently, researchers highlight that the presence of signal crayfish may affect river bed stability by modifying the microtopography and grain-grain fabric of gravel substrates which can significantly affect bed stability during subsequent flood events. Additionally, Gran and Paola used a series of physical modelling experiments to study the influence of riparian vegetation upon river morphology and braided stream dynamics. Furthermore, Tal and Paola conducted laboratory experiments to demonstrate that riparian vegetation can cause a braided channel to maintain a dynamic and single-threaded channel. In these studies, physical modelling allowed input variables, such as water discharge, sediment discharge and grain size to remain constant between runs, while vegetation density of alfalfa sprouts was varied between runs, confirming that vegetation acts to increase bank stability and reduce the number of active channels.

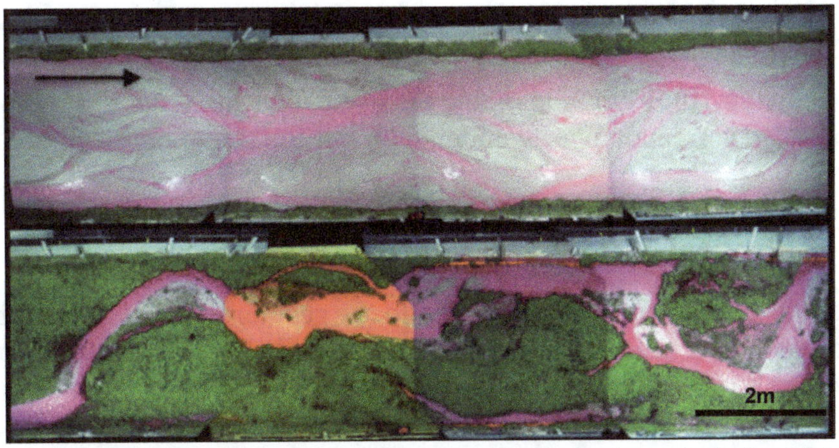

Transition from an un-vegetated braided channel to a dominant single-threaded channel with a vegetated floodplain in an experimental flume experiment.

Table: Summary of the advantages and disadvantages of using physical modelling and experimental laboratory techniques in geomorphological research framework.

Physical models advantages	Difficulties associated with using physical models
Incorporation of the appropriate physical processes without simplification or assumption. Allows the reproduction of complex physical phenomena.	Potential scale effects associated with simulating model variables in incorrect ratios. Consideration must be taken during planning stages.
Experimental control within a closed system allows rapid multi-variant analysis and testing of multiple variables. Ability to exclude extrinsic parameters.	Laboratory/model effects. Factors may be misrepresented/incorrectly reproduced after simulation in a laboratory environment.

Data collected simultaneously and with relative ease over large spatiotemporal scales once model is constructed and calibrated.	Exclusion/neglect of important functions and conditions which may have been overlooked or deemed to be insignificant by the experimenter.
Large degree of experimental control allows easy simulation of infrequent or hypothetical environmental conditions which would be difficult to observe in nature.	Construction and running is potentially expensive, labour intensive and time consuming. May require appropriate and continued support and funding.
Allow instant visual feedback. Provides qualitative insight into physical processes occurring. Calibration may be assisted by visual prompts/direct contact with physical model.	Data extraction can be difficult due to measurement effects. Results obtained may not be up-scaled to real-world situations/directly extended beyond the physical model.
Natural non-linear feedbacks and uncertainty in physical systems which may not be fully understood may be represented and modelled.	Construction and application may require previous experience, understanding or specific expertise.
Can be combined with other techniques to create 'hybrid/composite models', or used to calibrate or inform numerical model functioning and understanding.	May require specialist facilities and/or a large amount of space. Space constraints/lack of equipment may hinder experimentation.
Well-established technique applied to range of research applications. Numerous measurement techniques available, e.g. particle image velocimetry, Acoustic Doppler Velocimetry, pressure sensors, digital photogrammetry, laser scanning etc.	Substitution of materials may be required to ensure correct scaling. Physical model may not be fully representative of actual physical system.
May have reduced costs associated with data collection when compared to field data collection if using existing facilities/equipment.	Simulation of variables or conditions may not be possible at reduced scale within a physical modelling environment.
Control over system variables and inputs, e.g. sediment, water, vegetation. Bridges what can be simulated in the field and modelled numerically.	Equifinality may result in a misinterpretation of the fundamental processes occurring.

These physical modelling studies demonstrated the role of biota as a significant geomorphic agent. Many aspects of this field remain poorly understood but the use of physical models is of critical importance.

Advantages of Physical Models

Physical models provide a number of advantages to the geomorphological user. The main advantages of physical modelling are associated with the controlled, closed environment in which experimentation can take place. Physical models allow rapid analyses of multiple variables with a large degree of experimental control – independent variables can be altered one at a time while dependent variables can remain constant to investigate cause and effect relationships and model responses to changing variables. Additionally, physical models allow the simulation and study of infrequent, hypothetical or large spatiotemporal scale scenarios. This is significant for events which may be impossible to observe or difficult to study in the field because of the long timescales involved, e.g. the influence of autogenic mechanisms on alluvial fan evolution. Furthermore, physical models allow complex physical phenomena (potentially not yet described or understood) to be simulated without requiring a mathematical or theoretical simplification of governing processes. This makes physical models an invaluable investigative tool to the geomorphological user.

Physical Model Limitations

Despite offering a number of advantages to the geomorphological user, physical modelling also has a number of shortcomings which the user must be aware of before any experimentation takes place. Firstly, laboratory effects due to the limitations associated with simulating natural phenomena under a simplified and scaled laboratory set-up may produce occurrences that are not present in natural systems. These may include cohesive and/or adhesive forces between molecules (e.g. clay or water) becoming greater than within a natural system. Additionally, scale effects, whereby fundamental phenomena are unable to be simulated in correct proportions to that of the physical system, may arise. These may render results misleading or incorrect. In addition, difficulties in extracting useful and transferable data from physical models may be encountered, whereby data obtained within the physical modelling environment cannot be up-scaled to real world scenarios. Problems associated with equifinality, where the same end state is reached through different processes and mechanisms may also be present within physical models. Furthermore, physical models are potentially difficult to validate and determine whether the model is performing adequately because multiple model runs are required to allow adjustment of model variables until the observed effects are comparable to those observed in nature. Validation is essential to ensure that a physical model performs relative to its real-world counterpart but is rarely considered in physical modelling. Despite this, problems with model validation are problematic in other modelling techniques.

Analogue Models

Analogue models are further abstract scale models. Analogue models engage radical changes in the medium of which the model is constructed. For the largest part commonly used analogue models are maps and remotely sensed images. On a map, the surface features of a landscape are reduced in scale and represented by symbols: rivers by lines, relief by contours and spot heights by points for instance. Remotely sensed images like aerial photographs and satellite imageries represent specific properties of the landscape systems at a reduced scale.

Static Analogue Model.

Maps and remotely sensed images are, apart from where a series of them is offered for different times, static analogue models. Dynamic analogue models may also be built.

Mathematical Model

A mathematical model can be defined as an abstraction in that it substitutes objects, forces, event etc., by an expression containing mathematical variables, parameters, and constants involving the acceptance of a number of idealizations of variety of phenomena studied and in assigning to the assortment of entities involved some austerely defined properties. The indispensable features of the phenomena are then equivalent to the relationship between firm abstract symbols, which we can write down. The phenomena in consideration bear a resemblance to closely something enormously basic, with very few attributes. The resemblance is so accurate that the equations are a kind of working model, from which we can forecast features of the real phenomena which we have never observed.

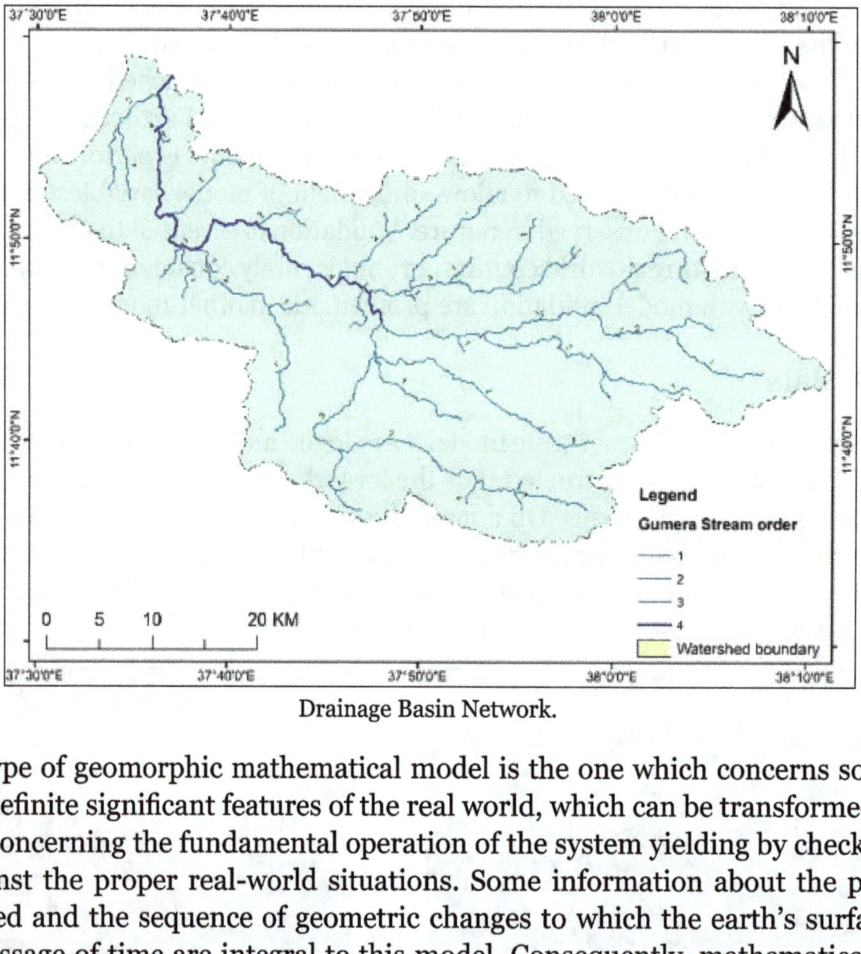

Drainage Basin Network.

The general type of geomorphic mathematical model is the one which concerns some simplified statement of definite significant features of the real world, which can be transformed according to assumptions concerning the fundamental operation of the system yielding by checking the model forecasts against the proper real-world situations. Some information about the primary mechanisms engaged and the sequence of geometric changes to which the earth's surface experiences with the passage of time are integral to this model. Consequently, mathematical examination follows the symbolic statement of the assumed primary static and dynamic relationships and the logical mathematical conclusions then are tested against the real world and the effects envisaged by the model point toward the success which we have had in building the original model. The dissimilarities exposed then might lead to the unearthing of auxiliary causes and these observed facts may be understood in greater detail. Mathematical models are frequently divided into categories like deterministic and stochastic.

Those based on classical mathematical notions of exactly predictable relationships between independent and dependent variables(that is, between cause and effect) are called Deterministic mathematical models and comprise of a set of exactly specified mathematical assertions from which exceptional consequences can be deduced by logical mathematical arguments. Such models are very closely concerned with relationships and driving forces between the factors identified in the simplified model. The most widespread type of deterministic geomorphic model engages the transformation of slope profiles under a variety of assumptions.

Stochastic mathematical models include random components that permits different possible results emanating from given set of initial conditions. An example of stochastic model is one in which simulation of individual streams originating under varied conditions or forces operating at random link forms stream network.

Numerical Modeling

Numerical models seek to represent the interaction between landscape forms and processes through mathematical equations. By integrating these equations over space and time, numerical models have allowed geomorphologists to extend enquiry beyond observation alone, and explore landscape dynamics over a range of temporal and spatial scales. Choosing the correct temporal and spatial scale of investigation, the correct processes that control landscape form at these scales, and then converting this conceptual model to a mathematical representation of these process-form interactions is not straightforward. The decision requires careful consideration of process dominance and scale, the ability of equations to parameterise these processes, computational resources, and data availability to constrain model parameters and evaluate model performance.

Numerical models provide some liberation from the temporal and spatial shackles imposed by enquiry through observation alone; models have allowed exploration of landscape processes and evolution over spatial scales ranging from particles to plate tectonics and temporal scales ranging from milliseconds to millennia. Numerical models have shown potential as powerful tools for understanding reductionist process-form interactions and also the relative importance of autogenic versus allogenic controls for larger scale system behaviour.

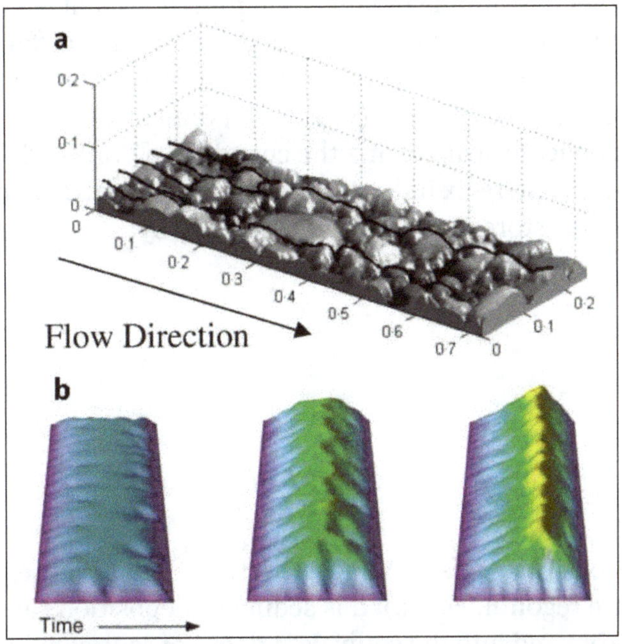

Contrasting scales of model application: grain scale predictions of particle paths (black lines) over water worked gravel, convergent orogen formation modelling, considering techtonic uplift and surface erosion.

In order to fully exploit the potential for numerical models to elicit understanding of form-process interactions, and inform landscape management, numerical models must be developed and applied carefully considering three key questions:

- What are the relevant form-process interactions at the scale of enquiry?
- What are the correct mathematical representations of these processes?
- Are there appropriate data to constrain model parameters and evaluate model predictions?

The strength of the assumptions made in developing and applying a model will determine the validity of model predictions, the strength of conclusions derived from model application, and therefore the ability of models to inform us of real world process form phenomena. A Bayesian approach emphasising iterative dialogue between model development and data collection is recommended as a robust means to appropriately develop numerical models, and therefore geomorphic understanding.

Model Structure

A model M contains equations with associated parameters θ that represent the functional relationship between a vector of driving conditions D (e.g. rainfall), a vector of initial system states X_0 (e.g. landscape elevation), and vectors (with length t, the length of the simulation) representing future system states X, and outputs Y (e.g. catchment runoff/sediment flux):

$$Y,X = M(\theta, X_0, D)$$

Geomorphic models are generally concerned with the action of a number of processes which locally transport mass (e.g. sediment, including organics and nutrients) that lead to changes in landscape form (X) at a specific point over time (t). Though models generally differ in the processes evoked to move sediment, a fundamental approach governing most geomorphic models is to divide the landscape into units called control volumes (in 1, 2 or 3 dimensions). In most models a quasi-2D conservation of mass is applied by calculating the change in elevation d- in response to sediment flux into and out of a control, where dx indicates the size of the control in one dimension. The control may be divided into three stores, and M partly specified by:

$$\frac{\partial H_r}{\partial t} = T_u - S_c$$

$$\frac{\partial H_s}{\partial t} = d - \varepsilon + S_c$$

$$\frac{\partial H_t}{\partial t} = -\frac{\partial ql}{\partial x} + \varepsilon - d$$

where H_r is depth of bedrock (m), T_u is tectonic uplift (m), S_c (ms^{-1}) represents the rate of conversion of rock to soil/surface regolith, H_s (m); d is sediment deposition rate (ms^{-1}) and -is sediment entrainment rate (ms^{-1}) from and into the equivalent depth of sediment in transport H_t (m), and q_s (m^2s^{-1}) is sediment discharge across the surface, dx. Up to specifying the source terms, accounting for density/particle size differences between the stores, and developing an appropriate numerical solution, Equations above can generally be used to simulate the evolution of any point in the landscape, though some specific exceptions apply.

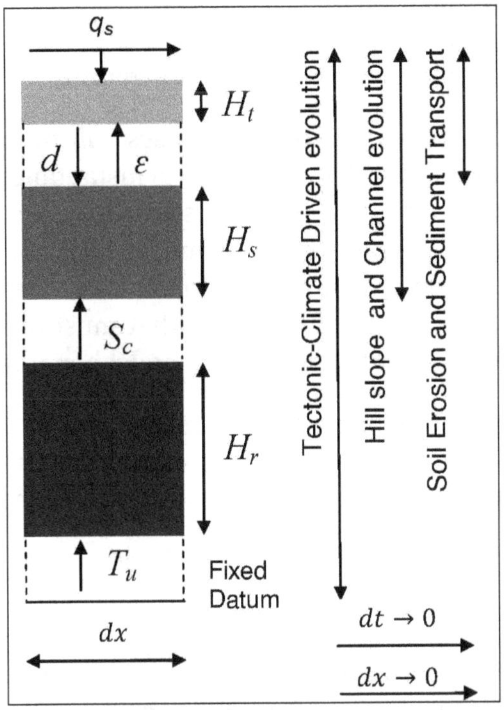

Key fluxes governing sediment movement within a one-dimensional control volume. The temporal scale of model application determines the relevant equations and methods of parameterisation.

To completely specify M, first a conceptual understanding of the relevant source terms on the right hand side of equations and the processes that control them are required at the scale of enquiry. Second, the conceptual model needs to be codified into a set of equations, and an appropriate analytical/numerical solution sought.

Process Dominance and Scale

The processes that govern changes in H_r operate over larger timescales (e.g. are relatively slower) than the processes governing changes in H_s which in turn are slower than processes governing changes in H_t. Therefore as the timescale over which a model needs to be applied tends to zero, so the number of relevant stores and processes that control d- also reduce.

In catchment systems over millennia climate induced fluctuations in sediment transport and rock breakdown, alongside tectonic uplift, govern the evolution of plate tectonics. Therefore, changes in H_r are important and need to be included in the conceptual model of landscape change. Over decadal and centennial timescales, catchments predominantly evolve in response to climate-induced fluctuations in sediment transport, and therefore the stores of sediment in the landscape evolve through transport between control volumes, and H_r may be considered fixed and the processes that control them (T_u and S_c) relaxed. At this scale the state vector (X) needs to consider not only the evolution of sediment and mass, but also controlling and interacting factors such as vegetation and potentially anthropogenic influence. Alongside exerting control on sediment flux directly, these factors will also respond independently to climatic changes, creating potentially complex, and non-linear landscape feedbacks. At even smaller temporal scales, sediment flux is controlled by current weather conditions and the effect of previous events operating at all scales which manifest their effects through the model initial conditions. Therefore many system states are fixed (e.g. H_r and vegetation cover) and only need to be specified in the initial conditions (X_o), with no additional equations required to simulate their evolution.

Similarly, over different spatial scales different processes will become important in controlling landscape behaviour. At small spatial scales on hill slopes instantaneous fluxes of water (raindrops and overland flow) control grain scale movements of sediment. As catchment size increases overland flow concentrates to form rills, gullies and channels which, alongside mass movements, are increasingly important in controlling catchment sediment flux. Therefore as the spatial and temporal scales of interest reduce, the range of processes that must be considered also reduces, and other, larger scale processes are manifest through the model boundary conditions (X_o).

Developing a model to address a specific problem therefore requires a sound conceptual understanding of the time and space scales over which processes operate to control the landforms in question. The extent to which a process is represented in a model, however, depends on its mathematical formulation.

Process Representation

Experimental work has been conducted, both in the laboratory and in the field, to investigate surface process, such as rain splash and overland flow driven erosion. Alongside fundamental, physical equations (e.g. Navier-Stokes equations of fluid motion), many such studies have provided us with predictive equations from which process form interactions can be simulated. When combined with conservation of mass equations and integrated over time, such models represent the fundamental mechanisms by which climatic fluctuations manifest in catchment-scale landscape evolution.

Given the temporal and spatial constraints on observation, the majority of experimental work has attempted to parameterise the processes in equation, through what may be determined process-based models. However, even at small spatial and temporal (reductionist/observational) scales, different process parameterisations have been developed.

In catchment systems erosion, transport and deposition of sediment by water is controlled by both transport limited processes (T_L; e.g. presence and power/stress imparted by water at the surface) and supply/detachment limited processes (S_L; e.g. the resistive forces at the sediment bed that impede sediment movement). The most widely applied predictive equations have calculated a sediment transport rate (q_s) as a function of T_L and/or S_L, which implicitly assume sediment transport is in equilibrium (and potentially at some capacity), and evolve H_s according to sediment flux into and out of the cell support. Though such methods may provide useful predictions, changes in sediment transport and soil depth are inherently in disequilibrium. Alternative parameterisations have been developed that explicitly calculate an entrainment and deposition flux into, and out of transport H_t,. Developing better parameterisations of equation have been limited by the difficulty of measuring sediment in transport. Therefore, even at scales where monitoring can take place to parameterise system processes, competing process representations may be derived reflecting process uncertainty, and also different experimental setups.

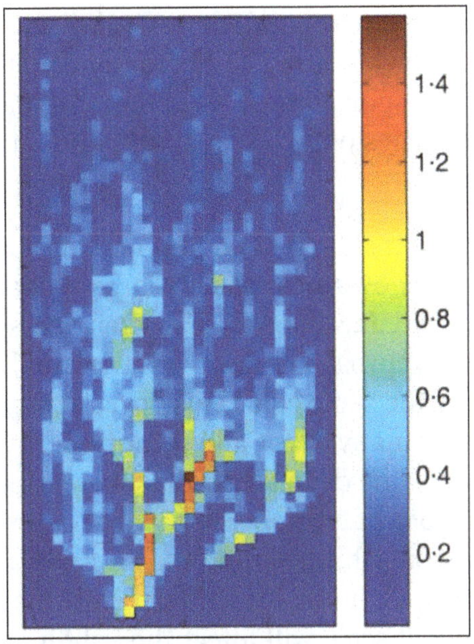

Spatial pattern of at-a-point total sediment movement (kg) predicted by MAHLERAN soil erosion model when applied to an 18×35m runoff plot.

Specifying the source terms in equation may also require separate models that simulate the behaviour of phenomena that control landscape change. In catchment systems transport limiting factors controlling sediment transport are primarily derived from free surface flow. The fundamental equations that simulate free surface flow are the 3D Navier-Stokes equations. Although it has been argued that flow sediment interactions should be modelled using these equations, various simplifications are typically made because of computational limitations involved in deriving accurate numerical solutions – a class of models often termed "reduced complexity" models.

In the case of hydraulic/hydrologic modelling, such simplifications include depth averaging to two dimensions and the fully dynamic 1D St Venant equations with further simplifications thereof that neglect potentially unimportant terms to derive the diffusive- and kinematic-wave models. Additional flow simplifications have led to a number of cellular approaches for flow routing, which employ simplified rules to route flow in river channels and hill slopes. Furthermore, some approaches have sought to employ simpler cellular and physically based rules to simulate morphological change, and have simulated sand dune formation meander migration, braided river evolution and floodplain evolution.

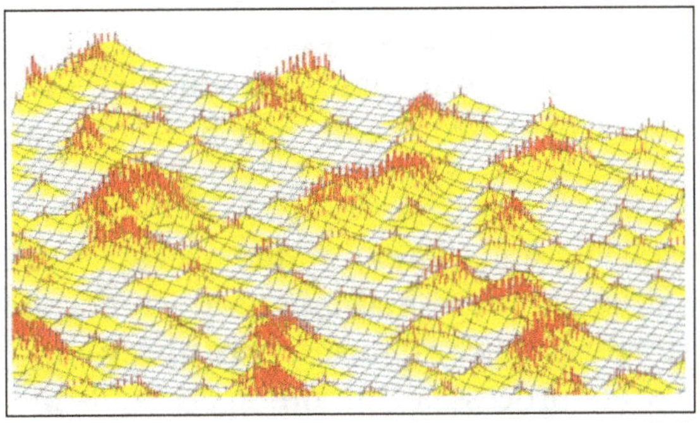

Nebkha dunes and vegetation (dark bars) simulated with a cellular approach.

The specific conditions under which such equations are 'valid' should be carefully considered (see Lane and Cao and Carling for a consideration of hydraulic equations in a geomorphic context). The scale and method by which transport limited conditions are modelled may be more important than the equations that link properties of flow to actual sediment movement, given the potential for non-linear error propagation.

The increased availability of high resolution topographic data has facilitated the application of small scale model parameterisations over increasingly large domains. For computational reasons and availability of other distributed data, however, it is often necessary to model at coarser resolutions in both space and time. One approach to deal with the problem of process parameterisation at coarser scales is simply to apply the same equations developed at smaller scales. However, a key problem with this approach is that many if not all geomorphic laws are scale dependent. For example many geomorphic laws are slope dependent, thus increasing model cell size reduces slope, and can lead to inaccurate predictions of erosion. Such inaccuracies occur because of how changing scale of resolution affects both geomorphic laws and the laws governing flow. Inferring the validity of a model equation independently of the grid within which it is applied may be difficult. Thus in model development changing model scale will affect the validity of all processes included in the model.

Although all spatial parameterisation is lumped to some degree, critical scales in the landscape that govern larger scale behaviour (e.g. the scale of interest) should be considered. For example, on hill slopes and in channels coarser scale models will fail to account for the spatial heterogeneity of flow. Given the relationship between sediment transport and flow is strong and non-linear, neglecting this heterogeneity will lead to an under-prediction of erosion and sediment transport. Relying on equilibrium concepts to model sub-grid scale channel features is a popular approach to deal with this problem in landscape evolution models, but relies on equilibrium concepts to model potentially non-equilibrium behaviour. Therefore the specific scale at which a model equation is a valid representation of sub grid-scale processes is an important consideration when developing a numerical model.

At larger temporal scales of enquiry models have to deal with the disparity between the timescales of individual events (e.g. rainfall runoff) and the evolution of landscapes. As a result of computational limitations, many landscape evolution models applied over larger temporal scales relate at-a-point discharge to upslope contributing area, which implies runoff is in equilibrium with uniform rainfall. Such approximations subsequently used to simulate sediment transport fail to account for spatial and temporal variability. Sub-scale events may be important in controlling runoff production and therefore the sediment transport that actually governs landscape behaviour. Simple averaging to an effective event may miss that some events are more important in controlling erosion and sediment transport than others. Furthermore, modification of the landscape by continued operation of smaller events may be (more) important in controlling morphological change.

The issue of what scales to consider and therefore what processes to resolve explicitly in a given model structure points towards a fundamental issue for geomorphologists: given natural systems often display nonlinear, threshold responses, it is uncertain - and debated in the literature - to what extent small scale processes (in both space and time) control larger scale system behaviour. It is therefore uncertain to what extent fine scale processes need to be resolved explicitly in geomorphic models, or whether simpler treatments, relying on for example regime theory, are applicable. Choosing a specific process representation therefore reflects a specific modelling hypothesis regarding the relevant processes governing a different problem. Multiple representations, and therefore hypotheses of the same processes may require investigation. To help overcome this issue, data are required to constrain model parameters and evaluate model hypothesis.

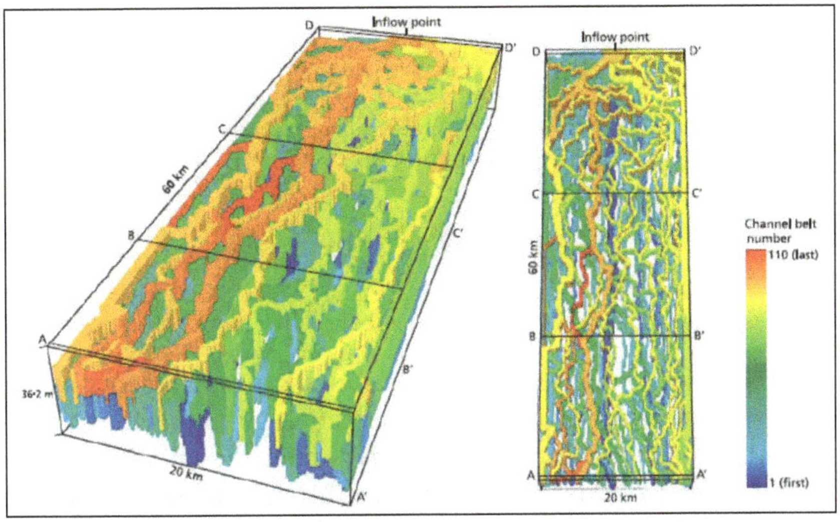

Simulated channel belt and floodplain evolution (e.g. by bifurcation, avulsion and aggradation) in response to base level rise.

Data and Model Evaluation

Data availability is an essential factor governing model development, as data provides the modeller with the ability to constrain model parameters and evaluate the quality of model predictions. The evaluation of model process representation is an essential step as it often occurs prior to application of models to investigate so called "what if" questions. Such model application is often at space and time scales over which data are insufficient to differentiate between competing model hypothesis.

In the ideal case, all process parameterisations of equations will be known and data will be available at the scale of the model cell size to constrain model parameters and the initial model conditions. For example in soil erosion modelling distributed information on particle size may be required to parameterise both roughness for overland flow modelling and the supply limiting factors controlling sediment entrainment. However, often such data are unavailable, or inconsistent with the scale of model application.

As a consequence, models may be calibrated by comparing model outputs to observations. In the case of catchment modelling, models are typically calibrated by adjusting internal model parameters to derive the best fit between model outputs (e.g. sediment/water flux at a catchment outlet) and the equivalent observations at specific locations. In morphological modelling, distributed model predictions may be compared to: observations or morphological change to results derived from models with a stronger physical basis for prediction and also to results derived from physical model experiments.

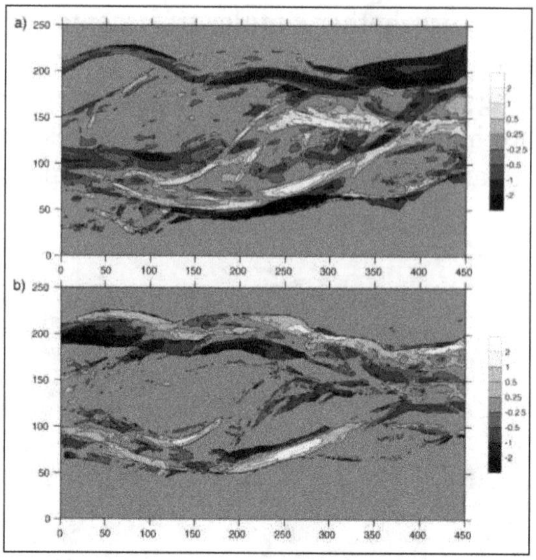

Comparison of measured (a) and modelled (b) elevation change in the braided Avoca River, New Zealand.

A number of implicit assumptions made in model calibration may potentially undermine model application; First, it is often assumed that parameter uncertainty is the only form of modelling uncertainty; Second, that the model is equal to reality; Third, the initial states are the true initial states (e.g. the DEM is error-free) and Fourth, that the input/driving conditions and output data to evaluate model performance are true. In most, if not all situations these conditions do not hold due to the problems of modelling an open system, where true model validation and verification is impossible.

As a result of these assumptions incorrect model parameters can be identified that reproduce catchment outlet conditions (i.e. larger scale measurements) with insufficient consideration of how well they reproduce the internal spatial patterns of process-form interaction that ultimately control larger and longer scale response. Furthermore, model parameters may be identified that are highly unique to specific settings and inapplicable elsewhere because of the non-linear open nature of natural systems. Worse still, a number of parameter combinations within a specific model may provide equally good predictions - a form of model equifinality. Similarly, another form of model equifinality may occur if the data are insufficient to differentiate between competing models. Equifinality has arisen at a range of scales, from using metrics of landscape form to differentiate between transport and supply limited models of landscape evolution evaluating alluvial fan evolution and at smaller scales when applying models with complex, and ill constrained parameters. However, equifinality is not all bad if it avoids over confidence in the information content of data and therefore the potential rejection of good model structures. In the face of equifinality, simpler models may be preferred that are justified by the data. A model is only as good as the data available to constrain model structure, parameters, and therefore predictions.

Accounting for Uncertainty when Developing Numerical Models

The preceding sections have discussed the main issues to consider in developing and applying a numerical model to address a geomorphic problem. Even if we consider our conceptual model of the system and the dominant processes to be accurate, computational resources and data availability will limit our ability to apply the preferred model at the desired scale. Furthermore, uncertainty surrounding the relationship between process dominance and scale is a fundamental geomorphic question governing model development. Therefore there is no single answer to the three questions posed in the introduction, and nor can each be answered independently. As a result of these factors a model can and will only remain as a (working) hypotheses of how processes and landforms interact. In order to develop, evaluate and use models in geomorphic enquiry, uncertainty in both data and models needs to be dealt with in a robust manner.

Considering numerical modelling from a Bayesian perspective provides a suitable framework for robust model development of non-linear open systems. As a result of modelling uncertainties we should be interested in obtaining the probability P() that the model (M) and its parameters (θ)are a correct representation of reality, given the available data (Y) initial model conditions (X_0)and driving conditions(D). In order to do this we combine our prior beliefs about the model structure P(M) and associated parameters P(θ|M) (which are dependent on the specific model structure) with some data using a Likelihood Function (e.g. a measure of model performance based on a given dataset) $P(Y|\theta,M,X_{0,}D)$ to obtain our posterior belief from Bayes' equation.

$$P(Y|\theta,M,X_{0,}D)$$

$$\propto P(Y|\theta,M,X_{0,}D)P(\theta|M)P(M)$$

Thus, our confidence in the model is specifically dependent on the data to constrain initial conditions, driving conditions and that used to evaluate model performance. In many applications only a single model structure is considered, and associated parameters are either derived from previous studies or are optimised to the specific data available. As a result the final two terms in Equation above collapse to a single set of structural assumptions. In such cases overconfidence in the data for the reasons discussed above – both in its accuracy and its general applicability to a wide range of settings – may lead to inappropriate model rejection and narrowing of the posterior probability of all possible models. This may lead to an entrenchment of modelling concepts that may prevent wider exploration of P(M) and therefore potentially more appropriate models for particular circumstances.

A better position to develop models and therefore understanding of process-form interactions is to consider the uncertainty in different model parameters, and therefore consider a range of possible combinations of parameters P(θ|M) that may reproduce the data, according to a likelihood function that considers potential uncertainty in the data. Significant advances have been made in developing appropriate likelihood functions in the related discipline of hydrology that consider different forms of model uncertainty. Such statistical treatments of model uncertainty require further adaptation to geomorphic problems, including potential uncertainty in model boundary conditions and elevation data. Calibration and the related sensitivity analysis conducted by exploring adequately different parameter combinations that constitute P(θ|M) can guide the modeller as to which parameters are most important in controlling system response. Such information can be then used to guide further data collection targeted at constraining the most important parameters.

Furthermore, when different models P(M) and therefore different hypotheses of form process interactions are confronted with the same observations the strengths and weaknesses of different model structures may be identified. Such information can be used to guide further data collection and understand the conditions (e.g. Y,X_0,D) under which different process parameterisations (e.g. M,θ) are valid.

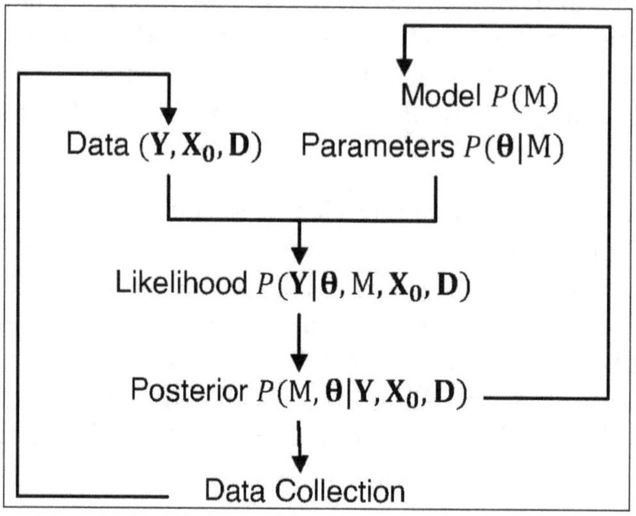

Schematic illustration of the iterative use of a Bayesian Approach for model development.

Therefore posterior understanding derived from comparison to data can guide further data collection. Depending on the similarities between the conditions used to derive the posterior (e.g. Y,X_0,D) and the newly collected data, the posterior in equation (e.g. the left hand side) can then become the prior for comparison to the newly collected data when combined with a likelihood function. When using model structures and parameters derived from previous applications it is up to the modeller to consider how appropriate such models are to the situation under current consideration, and whether such model application is supported by available data.

Bayes' equation therefore provides both a conceptual and probabilistic framework for model development that can explicitly consider uncertainty in models and data, and therefore appropriately frame the use of models in developing understanding of process-form interactions. Process-form understanding is developed iteratively through continuous dialogue between models and data. Furthermore such development provides a more robust grounding for subsequent model application to investigate "what if" type questions by considering a range of possible model structures supported by available data, which will prevent overconfidence in the results of a single model prediction.

Experimental Design Models

In terms of the contributions made to geomorphology, the foremost aspect of the physical systems approach engages the understanding that within a given range of observational data lies definite meaningful component parts that can be recognised by making use of a suitable experimental design. On the basis of the experimental design, constructed with reference to some conceptual model of the nature of the difficulty and appropriate operational definitions of its component parts, the numerical data are collected, checked with respect to their scalar and number-system distinctiveness and a 'data matrix' is composed. This data structure is frequently analysed by regression type statistical techniques to construct a simple variable system which workable correlations are acknowledged engaging the course and strength of assumed causation.

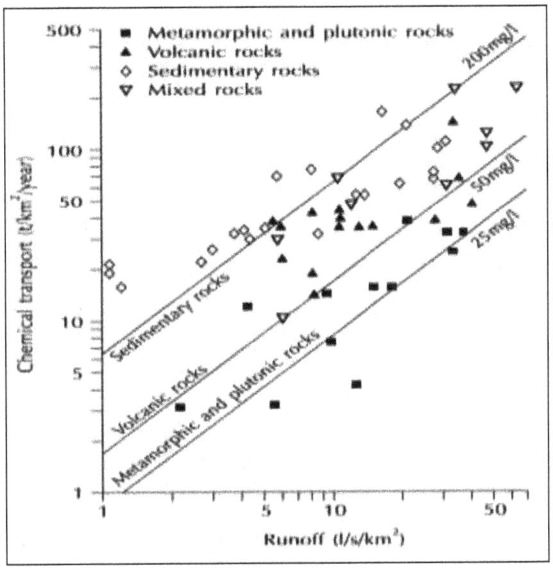

References

- Physical-Modelling: geomorphology.org.uk, Retrieved 04, August 2020
- Numerical-Modelling: geomorphology.org.uk, Retrieved 21, March 2020

Chapter 3

Understanding Primary Geomorphic Processes

Geomorphic processes refer to the endogenic and exogenic forces that cause physical stress and chemical actions to bring about changes in the configuration of the Earth's surface. Some of the different types of geomorphic processes are fluvial processes, glacial processes, hillslope processes, igneous processes and marine processes. This chapter discusses in detail these various types of geomorphic processes.

Aeolian Features and Processes

Aeolian processes, involving erosion, transportation, and deposition of sediment by the wind, occur in a variety of environments, including the coastal zone, cold and hot deserts, and agricultural fields. Common features of these environments are a sparse or nonexistent vegetation cover, a supply of fine sediment (clay, silt, and sand), and strong winds. Aeolian processes are responsible for the emission and/or mobilization of dust and the formation of areas of sand dunes. They largely depend on other geologic agents, such as rivers and waves, to supply sediment for transport.

Areas of sand dunes occur in inland and coastal settings, where they often provide a distinctive environment that provides habitats for endemic and rare or threatened species. In both coastal and inland settings, dune migration and sand encroachment may impact neighboring ecosystems and resources, as well as infrastructure.

Transport of fine sediment by wind may cause dust storms, events in which visibility is reduced to less than 1 km by blowing dust. Dust storms impact air quality in their immediate vicinity as well as in areas downwind. Deposition of dust may have a significant effect on the composition and nature of soils in arid regions and beyond. Far-traveled dust from distant sources may have a significant effect on soil chemistry and nutrient status.

Aeolian Environment

All land features formed by wind are also called 'Aeolian features' after the name of the Greek God of wind 'Aeolus'. Wind, although present everywhere, is a powerful agent of landform creation only in regions that are dry and mostly free of vegetation cover.

Wind erodes transports and deposits smaller particles like sand but may move even larger particles under special conditions. These processes lead to a variety of features. The scale of features ranges from very small to a vast expanse of several thousand square kilometers. The main features are hollows and grooves on rocky surface and mounds of sand. They have different shapes and sizes and are named and classified as such. Arid, semi-arid and sub-humid climatic regions are best suited for aeolian features. These drier parts occupy central and western regions of the continents that are on the leeward side of easterly rain bearing winds coming from the oceans.

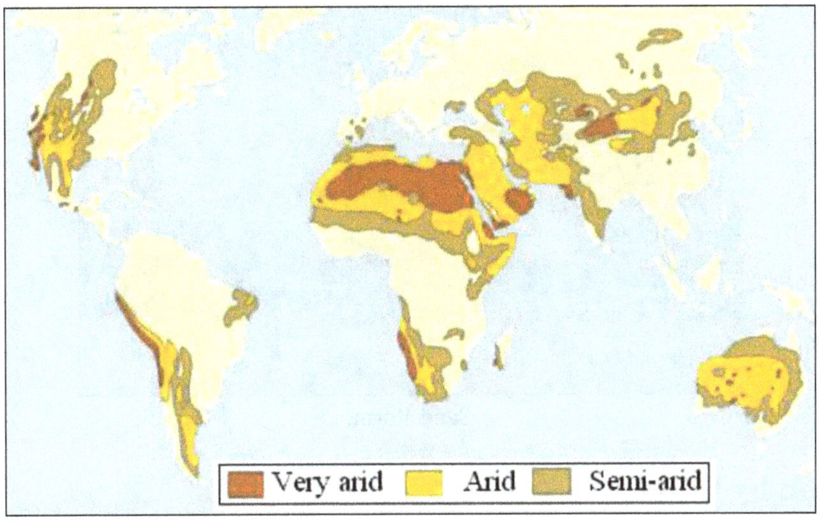

Distribution of aeolian environments in the world.

Processes of Wind Action

The three main processes that play vital role in development of aeolian landforms are Erosion by wind, Transportation by wind and Deposition by wind.

Erosion by Wind

Wind erosion follows two different processes:

- Abrasion: When strong winds blow and carry hard sand particles with them they attack rocky surfaces in their path of flow. This is known as 'sandblasting'. According to Bagnold, wind can lift sand particles only up to a height of 45 centimetres or 2 metres at the maximum. As such, abrasion by sandblast is limited to a zone close to the surface. Blackwelder has classified three different impacts of abrasion – i) polishing and pitting, ii) grooving and iii) shaping and cutting smooth surface on the windward side of the rock, also called faceting.

- Deflation: The process of lifting dry and loose particles and carrying them away is called deflation. Both the size of the particles and wind velocity play an important role in deflation. According to Richard Huggett, particles with 100 micrometre diameter are most suited for deflation. Particles larger than this cannot be lifted and carried unless the wind is especially strong. Deflation transports sand only to a limited distance. On the other hand, fine particles show a tendency to stick to each other and are not easily carried away by wind deflation. But once the finer particles are picked up they can be lifted up vertically higher up, kept afloat and transported to great distances. The process of deflation leaves hollowed features in its wake. Dust storms are another evidence of wind deflation.

Sand storm.

Transportation by Wind

Once loose particles are picked up by wind they are transported away by four different processes: a) creep, b) suspension, c) saltation and d) reptation. Each has a different role in landform creation.

- Creep: This mode of transportation involves movement of particles along the surface. Particles are pushed ahead as they roll and slip across the surface. This process influences larger grain size like sand and even pebbles that cannot be lifted but can be pushed by strong winds.

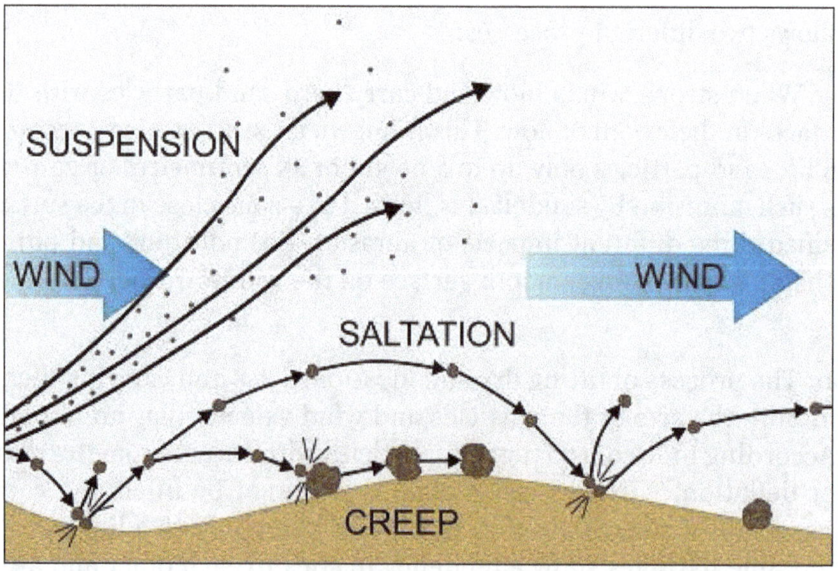

- Suspension: Smaller particles with a diameter of less than 100 micrometres are light in weight and can be lifted above the ground by wind eddies. Once these start floating they keep moving at a height until the force of wind totally dies out. Suspension may carry dust particles thousands of kilometres away from their point of origin.

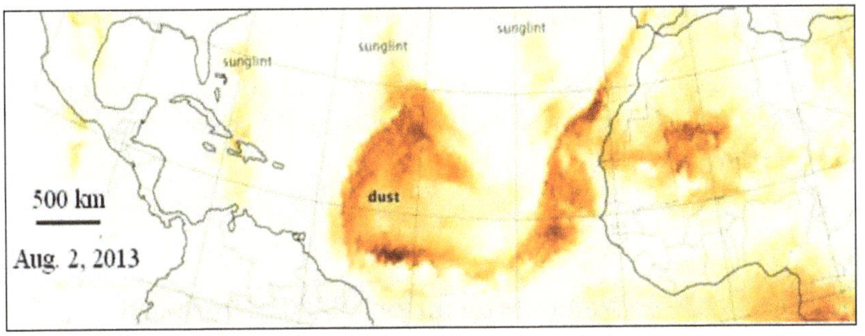

Satellite image of dust lifted from Africa and carried to South America.

- Saltation: Large grains of sand and gravel moves ahead in intermittent jumps. They are lifted and dropped repeatedly by gusts of wind. Maximum height for saltation is 2 metres. As they land and hit the ground they push other particles ahead. According to W. B. Sparks—in saltation "the impact derived from saltation is able to move grains six times the size of those forming the saltation. As a result, large particles are pushed by this process which cannot be moved otherwise by wind.
- Reptation: This process of wind transportation is closely related to saltation. As particles fall on the ground after jumping up, they create a splash and smaller particles are displaced in different directions.

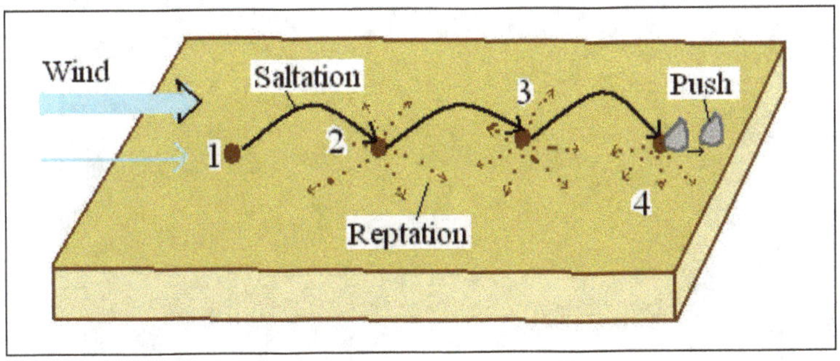

Deposition by Wind

When the velocity of wind is reduced, it loses its transporting capacity and drops the suspended fine particles carried by it, or stops creep and saltation started by it. The result of this inaction is deposition. On the basis of particle size there are two categories – fine particle deposition takes place far away from point of origin, while sand is deposited closer to the source region. Deposition may take place within the aeolian environments mentioned above or may be outside such areas. Four such areas have been mentioned by W. D. Thornbury—shore line, semi-arid river banks, extensive sandstone weathering areas and glacial out-wash zone. The deposited material forms various types of land features. Dunes of various types are the most conspicuous of all.

Types of Aeolian Land Features

The above mentioned processes are responsible for creating a variety of typical land forms, each indicative of the process responsible for its formation. On the basis of the two major actions of wind, aeolian landforms are classified into two broad categories — erosional and depositional aeolian features.

Aeolian Erosional Land Features

Wind erodes in two ways, one: it picks up lose particles and removes them to create depressions. Secondly, wind attacks rocks with sand particles and destroys weak rock beds. The following are the features formed by these actions.

Lag Deposit

While blowing over a surface, wind removes all unconsolidated fine particles. Those with less than 100 micrometres diameter are suspended and are taken to long distances. Those particles that are of 100 micrometres diameter, like sand, are removed gradually to short distances. The larger ones are left at their place of origin and keep rolling and shifting their place till they are tightly packed by this random jostling. These surfaces are called 'lag deposits', because the surface is made of particles that could not keep pace with the rest of smaller ones moving out and 'lagged behind'. They are also known as 'desert pavement' as the grains are fitted tightly, just like any man-made tiled pavement surface. The top of these desert pavements are polished by wind abrasion and have a thin shiny layer of oxides of iron and manganese, called desert varnish. These lag deposits have different names, e.g. desert armour in North America, serir and hammada in the Arab world and gibber in Australia.

Desert pavement, with desert varnish on the pebbles.

Deflation Hollow

As the name suggests, these are low-lying surfaces which have been cleared of all loose particles and converted into hollows. The size of these depressions may range from a few metres in diameters and depth, to several kilometers. The dimension, especially the depth, is controlled by the underground water table. As deepening reaches humid layers close to the water table, wind fails to move the moist particles and no further hollowing is allowed. These are also known as blowouts. Some of the examples of these deflation hollows are:

- P'ang Kiang Hollows – Several hollows are found in Mongolia, that are 8 kilometres wide and 60 to 120 metres deep. These have been described by C. P. Berkey and F. K. Morris as a work of wind deflation.

- Laramie Basin in Wyoming which is 14.5 kilometres long, 4.8 kilometres wide and about 46 metres deep.
- Quattara Depression in North Egypt, which has its deepest part 134 metres below the sea level.

The process involved in their formation essentially shows alternation of wet and dry periods. During the wet period, moisture helps in destruction of rocks due to agents of weathering. Unconsolidated rock grains created during this period are later transported by wind during the dry phase of the cycle. Repetition of the two phases gradually enlarges the depression.

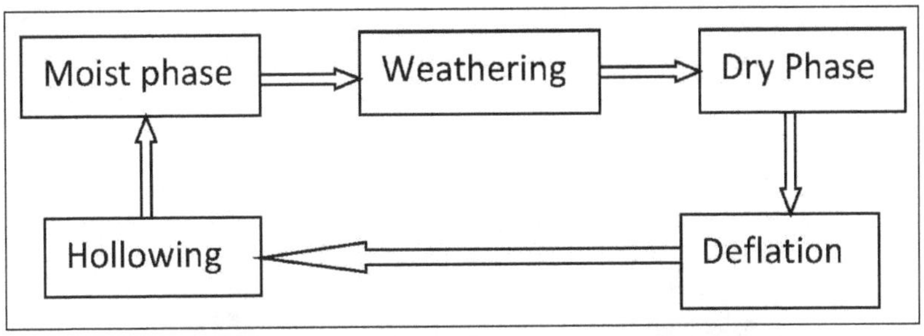

Robotic rover Curiosity has been on the planet Mars since August 6, 2012. On Dec. 5, 2015, it captured a view of the undisturbed surface of a Martian sand dune called "High Dune". It shows lag deposit of coarse grains, remaining on the surface after wind removal of smaller particles. The image covers an area 3.6 by 2.7 centimeters.

Yardang and Zeugen

Yardangs are elongated grooves, first described by Hedin in Turkestan. Eliot Blackwelder used the term 'Yardang' for these grooves in 1934. These are parallel ridges separated by parallel 'u'-shaped grooves, both developed in the direction of dominating wind flow in the region. They are called Mega-yardang when they are large in scale. In central Sahara and Egypt yardangs are 100 metres long and 1000 metres wide.

It is believed that formation of yardangs is initiated by some conditions that favour differential wind erosion. Some believe that initial depressions are made by the erosive action of water, and are later enlarged by wind deflation and abrasion. Arthur Bloom has given the example of such narrow parallel gullies in the cold desert of the central Andes. Here the gullies are cut along the joints that run parallel to the strong winds. Such initial grooving is mostly held responsible for evolution of yardangs everywhere. The feature itself is considered 'dynamic', that is, yardangs are both destroyed and made continuously.

Zeugen (singular Zeugen): These are similar to yardang, except they are smaller in scale and grooving is related to softer material alternated with more resistant rock beds.

Ventifacts

Ventifacts are rock pieces with smooth, sand-blasted facets pointing to the direction of dominant wind flow. Sometimes there may be several such facets developed on a rock, indicating varying wind direction. All facets in this type of case intersect along sharp and angular edges. If the rock piece has a pyramidal three-faceted shape it is called dreikanter.

Besides the above main landforms other features that result from near-surface abrasion by sand-laden winds are mesa, butte and mushroom rocks. These are formed when the base of a rock projection is eroded while the top is untouched by wind action, or is protected by some harder rock. Weak rock beds are eroded while harder beds stand out.

Aeolian Depositional Land Features

All particles transported by wind are ultimately dropped under two conditions – one, if the velocity of wind drops; or two, if the wind meets an obstacle in its path. Different circumstances lead to formation of different landforms. Most important of these are sand dunes. Besides these, 14 sand ripples and sand ridges cover vast areas. Bagnold has classified all depositional features into two broad classes on the basis of scale.

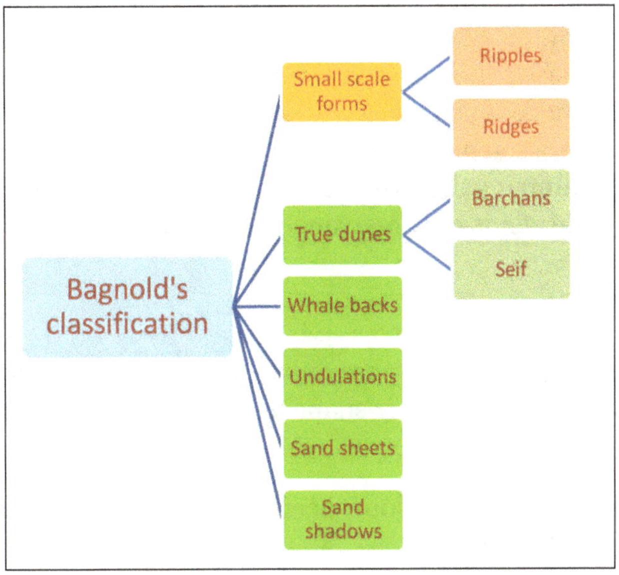

Sand Dunes

Sand dunes are defined as hills and mounds of sand. They have a large variety and are classified on different bases. Bagnold defines dunes as "mobile heap of sand whose existence is independent of either ground form or fixed wind obstruction". His classification mentioned only two types – barchans (transverse dune) and seif (longitudinal dune).

Dune Formation

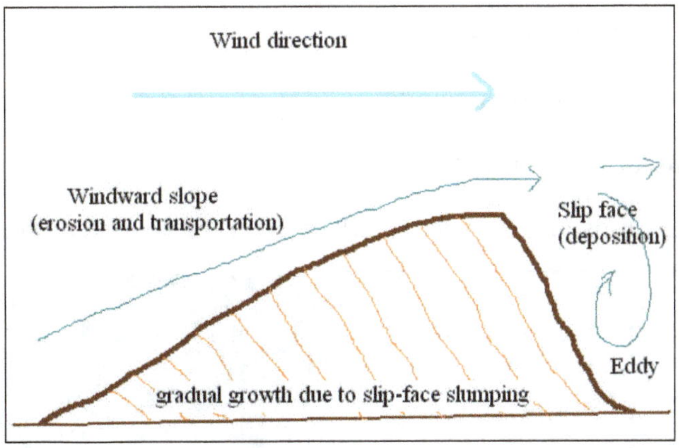

All dunes have more or less similar morphology and require similar ideal conditions. When wind meets an obstacle, it slows down and leaves some of the transported sand on the windward side of the obstacle. Gradually this deposition adds height to the evolving dune. When the crest of the deposit gains the maximum height possible under the available supply of sand and wind velocity, and the front becomes too steep, particles begin to slip forward and the leeward slip-face slope slumps. At this stage two simultaneous processes shape the dune. One, on the windward side sand particles arrive and move up the slope by creeping action; two, the leeward side keeps slumping and wind eddies remove lose particles from the dune. It is to be noted here that sand particles can achieve stability only on a surface with a slope of 34 degrees or less. As soon as the slope exceeds this critical point sand becomes unstable and shows a tendency to roll down. This critical angle is called the 'angle of repose' for sand, and plays an important role in all aeolian features. The processes active on windward and leeward sides continue, and gradually the dunes move towards the slip-face side. The balance between incoming sand on the windward side and the sand removed by eddies from the slip face, maintains the size of the dune as it moves ahead. Strong winds may add bulk to the dune, while gentle winds only rearrange its mass.

Barchan

Barchans are crescent shaped dunes. They are either single or may form groups. They migrate in the downwind direction, but maintain their shape as they move. They have the following distinct characters.

- Slip-face, it is a downwind steep slope, across which sand particles roll down when steepness is more than 34 degrees.
- Gently sloping windward slope, that receives fresh amount of sand. Sand here is compacted and not loose.
- A sharp lip marking the meeting line of the two slopes.
- Two limbs or horns gradually tapering and pointing downwind.
- They develop on rocky or lag deposit surfaces.
- Their height ranges from 0.5 to 100 metres.
- Their width ranges from 30 to 300 metres.
- They form under unidirectional winds.
- They can move at a rate of 40 metres per year.
- When several barchans join their horns they form transverse dunes.

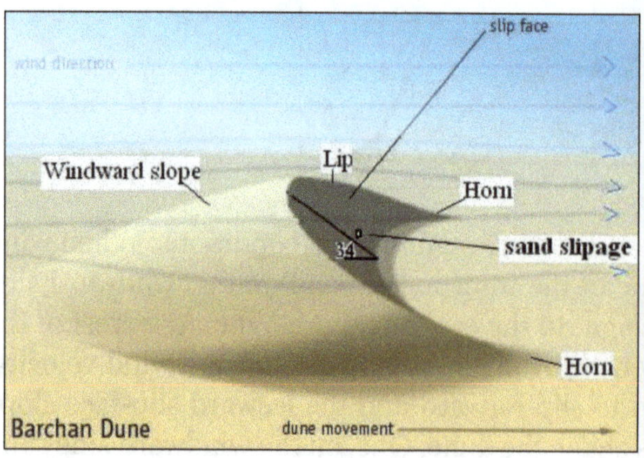

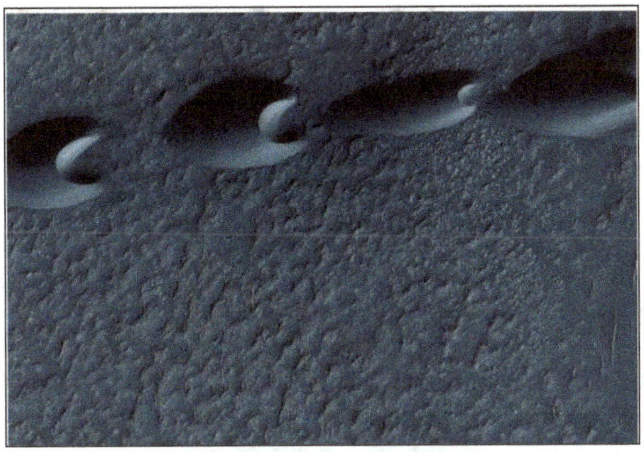

Isolated barchans on the Mars.

Seif

Seif is a linear dune with its axis aligned parallel to the prevailing strong winds. Its crest runs along its length and is marked by a sharp edge, hence it is also called sword dune.

There are several theories to explain it. One theory believes that seif is a result of bi-directional wind pattern. "The longitudinal or seif dune occurs when the wind regime is such that the strong winds blow from a quarter other than that of the general drift of sand caused by the more persistent gentle winds" – Bagnold.

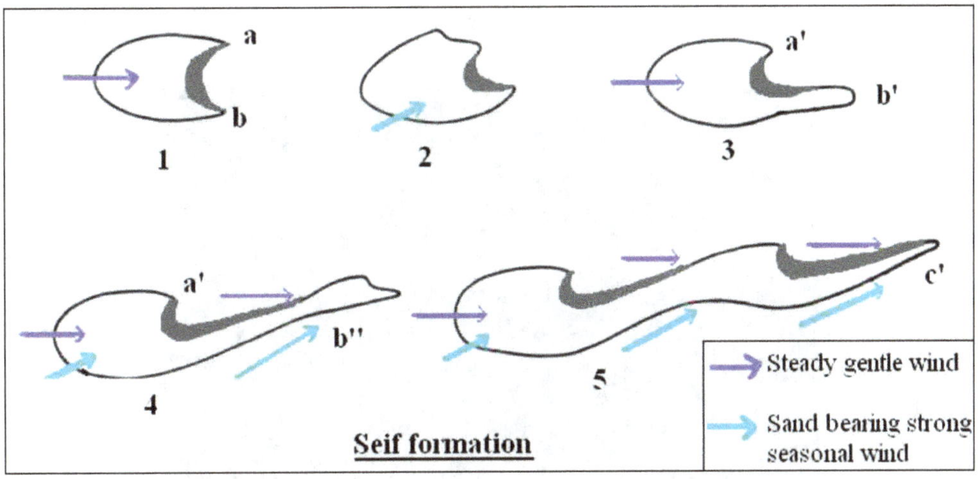

Seif formation

Seif is formed when gentle winds collect sand to form a normal, crescent-shaped dune, which is intermittently disturbed by strong winds, and the shape and mass of initial dune is modified. The two winds take turns to develop a seif in the following stages.

- Steady gentle wind forms a dune with two horns.
- When the seasonal strong wind approaches it adds a lot of new sand to the windward limb b and disturbs the balanced growth of the two horns.
- Again, during the gentle-wind phase, development and arrangement of all sand is aligned parallel to the wind flow. Thus b' is created.

- Alternatively, the seasonal wind again works on the windward side of the dune and adds more supply to b" and pushes it to the lee of its flow.
- The two winds work on the seif turn by turn, but the axis of the dune is controlled and kept parallel to its flow by the gentle wind, giving it a linear shape.
- Slip face on a seif is always to the lee of the prevailing wind, therefore it changes according to the changes in the wind direction. In general, the slip face runs on the two sides of the crest parallel to the axis of the dune.

Seif dunes have several summits; their number depends on the height of the seif. In lower seifs summits may be as close as 20 metres, while on higher seifs summits may be 500 metres apart. These dunes are capable of maintaining their straight alignment to winds for several kilometres. They can run across low cliffs or moderate depressions without losing their straightness. Southern Iran has examples of the highest seif dunes. Here, from base to crest they reach a maximum height of 210 metres. Their width usually is six times their height.

There are other explanations for linear dunes that develop parallel to the wind direction. According to one theory if the surface has some pre-existing linear features, like bands of raised resistant rocks or hills, then sand deposition is controlled by them to give shape to linear dunes. In such cases, the wind sweeps sand from the lower areas and move all sand particles towards the raised features. This movement of wind is a result of hot air currents that tend to move from the central lower zone to the bordering higher features and rise up along them. The winds form a circular cell between the hills and maintain the mass and shape of the linear dunes.

Linear dunes

Other Important Types of Dunes

- Parabolic dunes: These are typical of moist regions like the sea shore. In these dunes, the horns of the crescent are fixed, because shallow sand in horns allows vegetation growth, which stabilizes the sand. The higher, dry, central part of the dune keeps moving forward in the downwind direction. The shape of parabolic dune looks like an inversion of barchans, as its horns point towards the windward direction.

- Star dune: These dunes are formed in multi-directional wind regions. They have several limbs joined along a crest. Star dunes are fixed and have been at one site for several years.

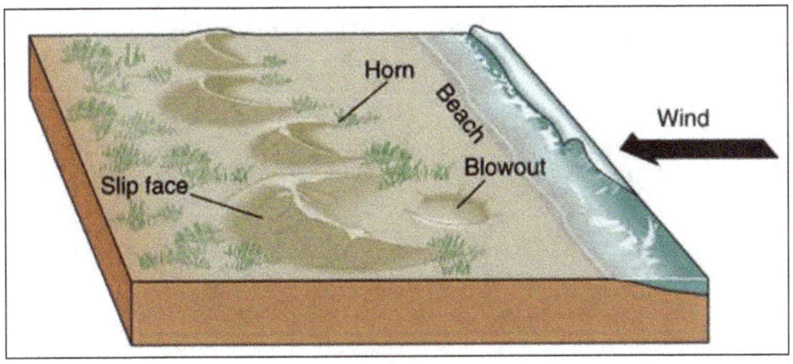

Parabolic dunes.

Star dune.

Sand Ripples

Ripples are small-scale aeolian features. They are 1 to 30 centimetres high and a few centimetres to some metres apart. The develop perpendicular to the wind direction. Their shape changes very quickly.

Ripples.

The figure is taken on Dec. 11, 2015 the rippled surface of the first Martian sand dune ever studied up close fills this view of "High Dune" from the Mast Camera (Mastcam) on NASA's Curiosity rover. This site is part of the "Bagnold Dunes". The dunes are active, migrating up to about one yard or meter per year.

Sand Ridges

Ridges are long, undulating aeolian features parallel to the wind direction. The main process responsible for their formation is saltation. In the beginning there are windward and slip-face activities, just like in the formation of sand dunes. Particles on the leeward side are protected from impact of the wind. The depression continually gets deeper as particles are removed from here rapidly. Due to saltation, large grains are pushed up along the windward slope to the crest of feature. The crest receives grains faster than it loses them. On the other hand, depressions lose grains faster than they receive, and hence get hollowed.

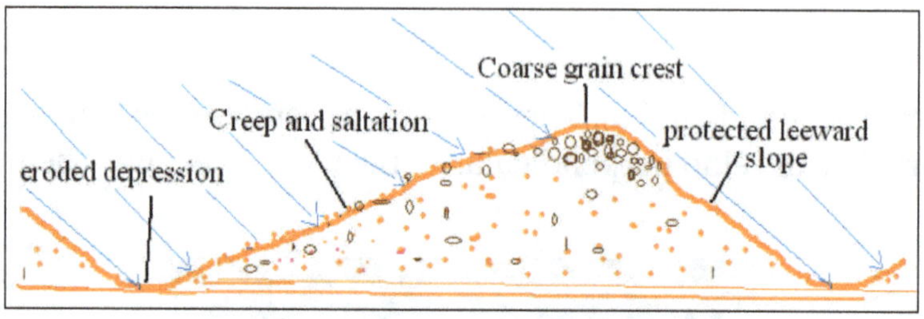

The above list of features includes only the most common, spectacular and widely discussed ones. There are other important aeolian land forms besides these, which should be mentioned here.

Whalebacks, Dunefields and Sand Sea

In the Sahara desert vast sandy areas are called ergs. These areas that have a level surface made of coarse grains are called 'plinth' by Bagnold. These are remnants of old dunes and seifs that have migrated from here. Whalebacks are "Coarsegrained residues or platforms built up and left behind by the passage of a long-continued succession of seif dunes along the same path".

Chains of transverse and seif dunes, barchans and other small scale features develop on these whalebacks to make dunefields. In the Western Egypt, the sand sea extends for 600 kilometres.

Dunefield.

Sand Shadow

Formation of this feature dependents on the presence of an obstacle in the path of the wind. The velocity of wind dips in the lee of such an obstacle, while the flow circumventing the obstacle maintains its force. As a result, weak flow fails to remove any sand particle that arrives in the leeward side of the obstacle; this allows sand to collect and form a depositional feature called Sand Shadow of the obstacle. It is formed close to the obstacle in its shelter.

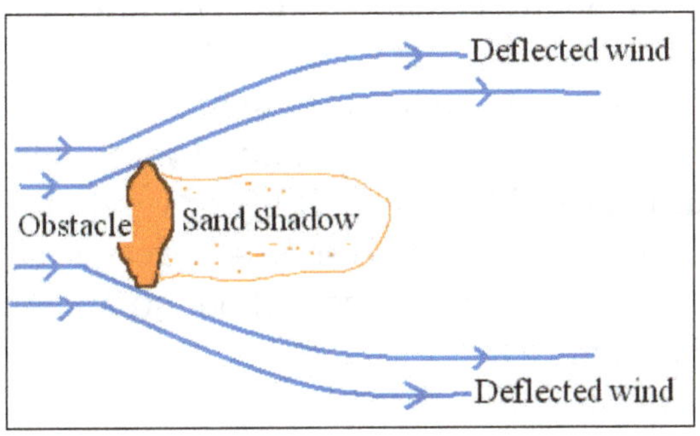

Sand Drift

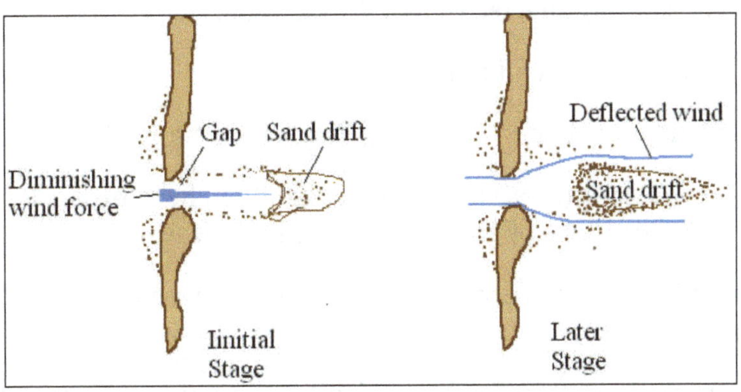

This feature is related to presence of gaps in landforms that allow wind to blow as a channelized strong stream. In such cases, the rest of the landform obstructs wind and transportation of sand while the gap allows unobstructed flow. All sand accumulating against such obstacles is directed to the gap and moves forward through it. Close to the gap there is no deposition because here the force of channelized wind is strong and transports its entire load. As wind moves farther from the gap and loses its force, it drops the sand it is carrying. Right in the line of the gap a mound builds up. Later this mound forces the wind to slow down and deposit more sand here.

Loess

Loess is very fine soil that wind has transported and deposited in thick layers far away from the place of its origin.

In 1949 Auguste Aubréville used the term 'desrtification' to describe his observations of areal expansion in the Saharan desert. In 1977 the United Nations Organization held a conference to discuss the problems caused by desertification and its remedies.

Desertification

"Although desertification can include the encroachment of sand dunes on land, it doesn't refer to the advance of deserts. Rather, it is the persistent degradation of dryland ecosystems by human activities — including unsustainable farming, mining, overgrazing and clearcutting of land — and by climate change" – UNCCD (United Nations Convention to Combat Desertification).

1. Causes for desertification: Degradation of land around the arid desert zone is a combined result of human, economic, political and climatic factors. Mostly, poverty is held responsible for a negative chain reaction in the delicate ecosystem of these regions. Overexploitation of soil through intensive farming, overgrazing, and destruction of natural vegetation for fodder and fuel are considered the main reasons. Once the soil loses its fertility or vegetation cover, the barren surface is exposed to wind action and its top may be lost. Also in some cases, adjoining desert sand is blown by wind and encroaches upon the non-desert areas. About 50 million people are threatened with displacement due to desertification.

2. Impact of desertification:

- 2.6 billion people depend directly on agriculture, but 52% of the land used for agriculture is moderately or severely affected by soil degradation.
- Land degradation affects 1.5 billion people globally.
- Loss of arable land is estimated at 30 to 35 times the historical rate.
- Due to drought and desertification each year 12 million hectares are lost (23 hectares/minute!), where 20 million tons of grain could have been grown.
- 74% of the poor (42% of the very and 32% of the moderately poor) are directly affected by land degradation globally.

3. Remedies for desertification: Sustainable agriculture practices, water conservation and rain harvesting are some suggestions to improve soil quality. Reforestation and checking overgrazing are solutions for prevention of soil erosion, as plant roots bind soil particles together. Tall tree lines are planted to break the wind speed and protect against encroaching sand from deserts.

Fluvial Processes

Fluvial geomorphology is the study of the landforms and physical features associated with river systems (including their channels and floodplains); and the sediment supply and transport processes that create them. Fluvial processes create a wide range of morphological forms that provide a variety of habitats within and around river channels. As a result, geomorphology is integral to river management. Fluvial geomorphology considers the processes of sediment transfer (erosion, transport and deposition) in river systems and also the relationships between channel forms and processes. The geomorphological form of a river channel and valley floor is influenced by many different factors and complex inter-related processes. Controls influencing river systems are both external (including catchment geology, topography, soil type, climatic trends and land management practices) and internal (including bed and bank materials, vegetation characteristics, gradient and flow conditions).

These variable controls and their interactions determine the character of fluvial processes, which in turn, influence individual channel forms and features. As an unmodified system, a river evolves in response to natural influences. However, rivers are often affected by human activities. Channel modifications including artificial structures alter flow and sediment movement, typically resulting in changes to river morphology (form), laterally (channel width, floodplain connectivity) and/or longitudinally (planform, bed gradient or depth). Changes in one part of the river catchment either through natural or human activity can result not only in geomorphological adjustment over time at that point, but also in changes upstream and downstream. An understanding of fluvial geomorphology adds value to the design of river modifications and structures (such as culverts, scour/bank protection and channel realignment) by identifying areas at risk of erosion and/or deposition. This leads to a potential reduction of maintenance costs by embedding mitigation to both protect assets and reduce/eliminate impacts on natural fluvial processes.

Streams and Drainages

Streams are flowing water bodies. The flow of water originates after every rain. Rain water falling on all the slopes of a land gets collected and start flowing on the land as overland flow. They merge together along the converging slopes, join together and run as a mainstream. The main stream, normally contain the cumulative volume of flow. The total water will be flowing out through a single outlet called as river mouth. The entire area encompassing the catchment zone up-to the river mouth, controlling this cumulative flow, is known as a river basin. It is also called as a drainage basin.

Stream Valleys

Every stream has its own valley. Valleys are common on the earth surface. A valley is a depression on the earth surface drained by, and whose form is changed by, water under the attractive force of gravity, between two adjacent uplands. Every valley has its own length and slope. The Horizontal distance measured in the cross section in a linear depression between two adjacent uplands is known as its length. The slope of a valley for a given reach refers to the difference in elevation between the upstream end to a point of reach along the downstream side.

Valley Development

Valley development is a unique hydraulic process. It happens due to three concomitant processes as:

- Valley deepening,
- Valley widening,
- Valley lengthening.

Valley Deepening

Valley deepening is effected by:

- Hydraulic action.
- Corrosion or abrasion of the floor of the valley.
- Pothole drilling along the valley floor and at the base of waterfalls.
- Corrosion.
- Weathering of the stream bed.

Valley Widening

Valley widening is accomplished by:

- Lateral erosion or planation by the stream in a valley.
- Hydraulic action.
- Corrosion.
- Rain wash and sheet wash.
- Gullying on valley sides.
- Weathering and mass-wasting.
- Role of incoming tributaries.

Valley Lengthening

Valley lengthening may take place due to three ways as:

- Extension by headward erosion.
- Increase in size of their meanders.
- Extension at the confluence due to subsidence.

The maximum downward limit for valley deepening is known as base level of erosion. In some cases, the channel length may increase due to a meandering course of a river, but the valley length is short.

Stages of a River

A river is said to be starting from an youthful stage, run though a mature stage and reaches its old stage depending upon its age, length of flow path, density of streams, population of its streams, discharge volumes and width of its mouth.

- Youthful Stage: A youthful stream is one that is undergoing initial development; consequently, it has a steep, irregular gradient, few tributaries or meanders, and a narrow floodplain. In addition, most of the erosion takes place in a downward direction. Youthful rivers are typically found in the highland or mountainous areas. They are characterized by steep slopes. They show a relatively small volume of water and rapid flow. Youthful streams may extend their source or head into higher elevation by cutting channels backward or upslope. Because their rapid flow leads to rapid erosion, youthful rivers tend to have narrow V-shaped valleys that are relatively straight.
- Mature Stage: Mature stage rivers possess a gentler slope and a flatter land surface. Mature rivers tend to have many well-developed branches streams called as tributarie. The river is characterized by a broad, flat river valley with a well-developed floodplain. The river channel will be relatively broad because the river, having reached a lower elevation and lost its steep slope, works on eroding the sides of the channel. They show a gently meandering course over the floodplain regions. A mature stage river normally reaches a stage of equilibrium in which its erosion and deposition are nearly in balance. Such rivers move a huge amount of water in their courses.

Power of the River

As time progress and erosion continues the stream gradient decreases. The number of course deflections, on the other hand, increases so that meanders begin to firm. Eventually, the degree of lateral erosion becomes equivalent to that of downward erosion. This intermediate stage of development is referred to as maturity.

Old Stage

During the old stage, the relief around the river becomes extremely flat, the river will have almost no slope and consequently very little momentum or speed. These are characterized by very elaborate and intricately meandering river courses. They show swampy areas, around the old rivers because the lack of slope leads to very poor drainage. The old rivers are very muddy. This is due to the consequence of large amounts of suspended material being carried by the river with low velocity. The river at this stage will show a high discharge of water. The final stage of stream evolution occurs when lateral erosion takes precedence over down cutting. These so-called old-age streams have very low gradient and wide floodplains.

Profiles of a Stream

Every stream has its longitudinal profile and lateral cross-sections. The longitudinal profile of a stream is analysed to know the water surface slope. It helps to compute the velocity of flow and the gradient. The transverse profile helps in identifying the nature of topographic floodplain and hydraulic floodplains. The actual water flowing zone in a river is called as the bankfull elevation.

The longitudinal profile of a river shows how a river's gradient changes as it flows from its source to its mouth. The long profile shows how, in the upper stage of a river's course, the river's gradient is steep but it gradually flattens out as the river erodes towards its base level. These are points where the gradient of the river changes suddenly and can be caused by landforms like waterfalls or lakes, where the lithology of the river changes and differential erosion takes place.

A River's Course

The course a river takes is split into three stages, the upper, middle and lower stage. In the upper stage, the river is close to its source and high above its base level. In the upper course, the gradient of the river is steep and the river is high above sea level giving it a large amount of gravitational potential energy that can be converted to kinetic energy later on. In the middle course, the river's gravitational potential energy gets converted to kinetic energy and the gradient begins to level out resulting in the river's velocity increasing. By the time the river reaches its lower stage, it has next to no gravitational potential energy but lots of kinetic energy resulting in a high velocity. In the lower stage, the river will be far away from its source, close to the mouth and not far above its base level. In the middle stage, it's somewhere in between. The total energy that a river possesses varies from one stage to another. It is because of changes in the river's height, gradient and speed.

Processes in the Upper Course

In the upper course, the river has a lot of gravitational potential energy. Hence, it has a lot of energy to erode vertically. The bed of the river is eroded greatly while the banks aren't eroded as much. The river at this stage mainly transports large pieces of angular rocks. This increases erosion processes. Vertical erosion is further increased by the rough nature of the channel in the upper course which increases the water's turbulence and its ability to erode. Erosion and transportation takes place, in large quantities, in the upper course when the river's discharge is high after periods of heavy precipitation. When the river's discharge falls, the river stops transporting the large boulders and deposits them.

Processes in the Middle Course

In the middle course, the river has a less gravitational potential energy but possess more kinetic energy. Due to this, the erosion shifts from vertical to lateral erosion. Corrasion is still the main erosive process as large particles are transported by saltation. The average load size has decreased in the middle course, so more load is being transported in suspension. In the middle course, the river can induce floods. There is also a possibility that the river may deposit its gravel and sand sized particles, onto its flood plain, in this stage.

Processes in the Lower Course

In the lower course, the river will have no gravitational potential energy. Hence, the erosion is almost exclusively lateral. There isn't much of erosion as the channel is smoother resulting in less turbulent flow. The main place where erosion takes place is where the river meanders. The average particle size is very small. The river's sediment load is mainly composed of silts and sclays. These are transported in suspension or even in solution. Similar to the middle course, when there is flooding in the river, it deposits all its load. Here, the deposition takes place at the mouth of the river, where the river meets the sea or a lake.

Erosional Work of Streams

Moving water erodes all earth's surface materials, not only along the bed but also along the sides of the channel. The river transports all the eroded materials on to a new location, and then deposits them. The materials deposited by a streams are called as alluvium. The ability of a stream to do all these works is a function of stream velocity and discharge. Running water is capable of conducting headward erosion, vertical erosion and lateral erosion.

A Three-Way Mechanism

Flowing water erodes in three ways. Firstly, the flowing water dissolves the soluble materials from the channel, and contributes to the stream's dissolved load or, solution load. Secondly, the impact of water, or hydraulic action, on the sides and bed of the channel, try to dislodge the materials and makes them available for transport, as a part of the stream load. Thirdly, the materials that are too heavy to carry in suspension, scoot and roll across the bed and eroding the channel by abrasion.

Rate of Erosion

The rate of erosion depends on many factors. Climatic factors include the amount and intensity of precipitation, the average temperature, as well as the typical temperature range, and seasonality, the wind speed, and finally, the storm frequency the erosion will continue till the base level is reached. The "Base Level" is the lowest (elevation) point to which a stream can flow and cut down. It is the ultimate level which is the maximum, and that is the sea level.

The Geological and Biological Factors

The geologic factors include the sediment or rock type, its porosity and permeability, the slope (gradient) of the land, and nature of structures likes tilts, faults, folds or weathered. The biological factors include ground cover, by vegetation or lack thereof, the type of organisms inhabiting the area, and the land use. Sediments consisting of more clay fractions tend to erode less than those with sand or silt fractions.

Types of Erosion

Types of Erosion carried out by running water are the following:

- Splash erosion,
- Sheet erosion,
- Rill erosion,
- Gully erosion,

- Stream erosion,
- Bank erosion.

Splash Erosion

Splash erosion is the first stage of the erosion process. Splash erosion is the detachment and air-borne movement of small soil particles caused by the impact of raindrops on soil. Splash erosion occurs when raindrops hit the bare soils. The explosive impact breaks up the soil aggregates so that all the individual soil particles are 'splashed' onto the soil surface. The splashed particles can rise as high as 60cm above the ground and move up to 1.5 metres from the point of impact. The particles block the spaces between soil aggregates, so that the soil forms a crust that reduces the infiltration rate and increases the runoff.

Sheet Erosion

Sheet erosion is the detachment of soil particles by raindrop impact and their removal downslope by water flowing overland as a sheet instead of in definite channels or rills. The impact of the raindrop breaks apart the soil aggregate. Particles of clay, silt and sand fill the soil pores and reduce infiltration.

After the surface pores are filled with sand, silt or clay, overland flow of water begins due to the lowering of infiltration rates. Once the rate of falling rain is faster than infiltration, runoff takes place. There are two stages of sheet erosion. The first is rain splash, in which soil particles are knocked into the air by raindrop impact. In the second stage, the loose particles are moved downslope by broad sheets of rapidly flowing water filled with sediment known as sheet floods. This stage of sheet erosion is generally produced by cloudbursts. The sheet floods commonly travel for short distances and may last only for a short period of time.

Rill Erosion

Rill erosion refers to the development of small, ephemeral concentrated flow paths, which function as both sediment source and sediment delivery systems for erosion on hill slopes. Flow depths in rills are typically on the order of a few centimeters or less and slopes may be quite steep. These are very different hydraulic environments. Eroding rills evolve morphologically with reference to both time and space. The rill bed surface changes as soil erodes, which in turn alters the hydraulics of the flow. The hydraulics of flow is the driving mechanism for the erosion process. The process of rill evolution involves a feedback loop between flow detachment, hydraulics, and the bed form.

The factors involved in rill erosion are:

- Flow velocity,
- Depth,
- Width,
- Hydraulic roughness,
- Local bed slope,

- Friction slope,
- Detachment rate.

Gully Erosion

Gully erosion is another type of erosion. It is also called as ephemeral gully erosion. It occurs when water flows in narrow channels during or immediately after heavy rains or melting of snow in mountainous regions. A gully is sufficiently deep that it would not be routinely destroyed by any tillage operations. The narrow channels, or gullies, may be of considerable depth, ranging from 0.3 to 0.6m to as much as 25 to 30m.

Stream Erosion

It is the Valley or Stream Erosion which creates a lot of landforms. The valley or stream erosion occurs with continued water flow along a linear surface. The erosion is both downward and headward. The downward action will be deepening the valley, and the headward erosion will be extending the valley into the hillside. In the earliest stage of stream erosion, the erosive activity is dominantly vertical. During this stage the valleys will have a typical V shaped cross-section and the stream gradient is relatively steep. When some base level is reached, the erosive activity switches to lateral erosion, which widens the valley floor and creates a narrow floodplain. The stream gradient becomes nearly flat, and lateral deposition of sediments becomes important as the stream meanders across the valley floor.

Bank Erosion

Bank erosion is the wearing away of the banks of a stream or river. Bank erosion is the wearing away of the banks of a stream or river. The roots of trees growing along a stream are undercut by such erosive actions. Bank erosion is a natural process. The natural process of riverbank erosion can produce favorable outcomes such as the formation of productive floodplains and alluvial terraces. The major causes of bank erosion are flooding, land use, stream management, over-clearing of catchment and stream bank vegetation, and poorly managed sand and gravel extraction.

Transportation Work of Streams

Once the surface material is detached from the source by the water, it can be transported by the water very easily. Fluvial transportation is the movement of earth's material enacted by the running water. As the particle size increases, the velocity needed to transport also increases. The material transported through the stream is called as it's stream load.

Processes Involved in Transportation

Transportation involves the following processes:

- Traction: It refers to the load that is rolled along the bed of the river.
- Saltation: It refers the load that is bounced along the bed of the river.
- Suspension: It refers to the load that is transported in a rivers' flow (current).

- Solution: It refers to the load that is dissolved by a river and then transported by it.
- Flotation: Refers to the material transported on the surface of a river.

Stream Load

Stream load is a geologic term referring to the solid masses carried away by a stream. The amount of solid load that a stream can carry, or stream capacity, is measured in metric tons per day, passing a given location. The stream capacity is dependent upon the stream's velocity, the amount of water flow, and the gradation. The stream load is composed of three components as the dissolved or solution load, the suspended load and the bed load. Bed load is that which is moved across the bed of the channel.

Bed load is transported in two ways:

- Traction, which is a scooting and rolling of particles along the bed.
- Saltation, a bouncing-like movement.

Saltation occurs when particles are suspended in the stream for a short distance after which they fall to the bed, dislodging particles from the bed.

Three Types of Loads

Water has the ability to dissolve soluble mineral substances on the surface and also within the groundwater system. The dissolved load comes primarily from groundwater seepage into the stream. Suspended load is comprised of sediment suspended and transported through the stream. Turbulent flow suspends clay and silt in the stream. Suspended load comes from material eroded from the surface bordering the channel and deposited in the stream, as well as, erosion of the channel itself. The stream capacity is the maximum load of sediment a stream can carry for a given discharge. The stream capacity increases with increasing flow velocity. Turbulent flow occurs under higher velocity thus increasing the water's ability to dislodge material from the bed or sides of the stream. Stream competence is the largest size material the stream can move under a given discharge.

Erosional Landforms

The erosional landforms of the running water are very wide. They are:

- River valleys,
- Canyons,
- Gorges,
- Waterfalls,
- Potholes,
- Ravines,
- Gullies and rills.

Canyons

A canyon or gorge is a deep ravine between pairs of escarpments or cliffs. It is most often carved from the landscape by the erosive activity of a river over geologic timescales. A canyon may refer to a rift between two mountain peaks. The word "Canyon" comes from the Spanish word *cañon*, which means "tube" or "pipe." The term "gorge" is often used to mean "canyon," but a gorge is almost always steeper and narrower than a canyon.

Rills

Rills are shallow channels eroded by threads of turbulent flow developed in the sheet flow. During rain storms rills erode headward on the steepest local gradient at faster rate. On open slopes they tend to form parallel to one another, converging in hillside hollows to form dendritic patterns. Rills are ephemeral features. They may be destroyed and recreated during major storms. The rills terminate at the base of slopes and thus they are not part of the regional drainage network.

Depositional Landforms

The depositional landforms created by the running water are:

- Floodplains,
- Alluvial fans,
- Deltas.

Floodplain

A floodplain is an area of land adjacent to a stream or a river course. It stretches from the banks of its channel to the base of the enclosing valley walls. This region normally experiences flooding during periods of high discharge. These are the surfaces of low relief developed on the alluvium adjacent to a stream. The flood plains become the stream bed during flood seasons. The Floodplains contain various features like:

- Point bars (lateral accretion),
- Overbank sediments (horizontal accretion),
- Levees,
- Levee crevasses,
- Splay deposit,
- Meanders,
- Neck cutoff,
- Oxbow lake.

Features of Deposition

The term alluvium is applied to all fluvial deposits with the exception of deltas. Streams carrying coarse sediments develop sand and gravel bars. These types of bars seen often in braided streams which are common in elevated areas. Bars develop in braided streams because of reductions in discharge. Two conditions often cause the reduction in discharge: reduction in the gradient of the stream and/or the reduction of flow after a precipitation event or spring melting of snow and ice. Sand Bars are the depositional features of a river. Point bars develop where stream flow is locally reduced because of friction and reduced water depth. In a meandering stream, point bars tend to be common on the inside of a channel bend.

Dunes and Ripples

Dunes and ripples are the primary sedimentary features in streams whose channel is composed mainly of sand and silt. Dunes are about 10 or more centimeters in height and are spaced a meter or more apart. Ripples are only a few centimeters in height and spacing, and are found in slow moving streams with fine textured beds.

Rosgen's Observations

Rosgen, proposed the following features that are created during the course of deposition by a river along its flow path:

- Creation of point bars.
- Point bars with few mid-channel bars.
- Numerous mid-channel bars.
- Side bars.
- Diagonal bars.
- Main-channel branching with numerous mid-bars and islands.
- Side bars and mid-channel bars with length exceeding 2-3 channel widths.
- Delta bars.

Channel Features

Streams with high sediment loads that encounter a sudden reduction in flow velocity generally have a braided channel type. In a braided stream, the main channel divides into a number of smaller, interlocking or braided channels. Braided channels tend to be wide and shallow because bed load materials are often coarse (sands and gravels) and non-cohesive.

Meanders

Meandering channels form where streams are flowing over a relatively flat landscape with a broad floodplain Technically, a stream is said to be meandering when the ratio of actual channel length to the straight line distance between two points on the stream channel is greater than 1.5. Channels in these streams are characteristically U-shaped and actively migrate over the extensive floodplain. Floodplains develop when streams over-top their levees spreading discharge and suspended sediments over the land surface during floods.

Levees

Levees are ridges found along the sides of the stream channel composed of sand or gravel. Levees are approximately one half to four times the channel width in diameter. Floodplain deposits can raise the elevation of the stream bed. This process is called aggradation.

Oxbow Lakes

Oxbow lakes are the abandoned channels created when meanders are cut off from the rest of the channel because of lateral stream erosion.

Alluvial Fans

An alluvial fan is a large fan-shaped deposit of sediment on which a braided stream flows over. Alluvial fans develop when streams carrying a heavy load reduce their velocity as they emerge from mountainous terrain to a nearly horizontal plain. The fan is created as braided streams shift across the surface of this feature depositing sediment and adjusting their course.

Deltas

Streams flowing into standing water normally create a delta. A delta is body of sediment that contains numerous horizontal and vertical layers. Deltas are created when the sediment load carried by a stream is deposited because of a sudden reduction in stream velocity. The surface of most deltas is marked by small shifting channels that carry water and sediments away from the main river channel. These small channels also act to distribute the stream's sediment load over the surface of the delta. Many deltas are triangular in shape.

Deposits of Deltas

Most deltas contain three different types of deposits as foreset, topset and bottomset beds. Foreset beds make up the main body of deltas. They are deposited at the outer edge of the delta at an angle of 5 to 25 degrees. Steeper angles develop in finer sediments. On top of the foreset beds are the nearly horizontal topset beds. These beds are of varying grain sizes and are formed from deposits of the small shifting channels found on the delta surface. In front and beneath the foreset beds are the bottomset beds. These beds are composed of fine silt and clay. Bottom set beds are formed when the finest material is carried out to sea by stream flow.

Glacial Processes

Glacial movement occurs when the growing ice mass becomes too heavy to maintain its rigid shape and begins to flow by plastic deformation. Glaciers and ice sheets are the world's main storehouse of fresh water. They are the thick and large masses of slow moving ice under the influence of gravity. They are formed by compression of snow. The mass of ice is mostly confined in the areas where the temperature is below zero degrees Celsius. They are found in the high latitudes or high altitudes. In higher latitudes, glaciers are seen even on the flat ground. In low latitudes, the glaciers are found in the higher altitudes. Ice accumulation is possible when there is greater amount of snowfall in comparison to snowmelt. Glacier moves downslope due to gravity. Therefore, slowly moving ice body known as glacier. Periglacial is referred to the areas at the periphery/edge of the glacier. It is characterized by low temperature. Its annual mean temperature is around zero degree Celsius. It summer season, the surface is free from ice but subsurface water is found in frozen state. Landscape is the surface of the earth as we observe while being in the field. It is basically the appearance of the land. The land itself is sculptured by many agents of erosion. Depending upon the processes of sculpturing, the landscapes are different. Glacial and periglacial landscapes refer to the land features developed due to the process of glaciations and periglaciation.

Glacial and Periglacial Conditions

Glacial condition exists where the temperature is always less than zero degree Celsius. Mostly precipitation occurs in the form of snowfall in glacial areas. The annual amount of snowmelt is less than or equal to annual snowfall over the area. If snowfall is more, the glaciated area is on rise; but when it is equal to the snowfall, it is stagnant. When the snowfall is less than the snowmelt, the glaciated area is on decline. Glaciated areas always bear the ice on the surface. It is not so only when the slope is too high to sustain the ice. The accumulated ice is on slide downward. In this way, it develops a large number of features.

Periglacial condition exists in the peripheral areas of the glaciated regions. In this zone, the annual average temperature is around zero degree Celsius. In winter, there is thin layer of ice and snow accumulation but it is not the same during summer days. Even during summer days, the subsurface moisture content is found to be in a frozen condition. In the soil zone, the temperature is less than zero degree Celsius. Freezing and thawing is a normal condition which is main cause of weathering here. Both glacial and periglacial conditions are quite different from each other.

Glaciated Areas

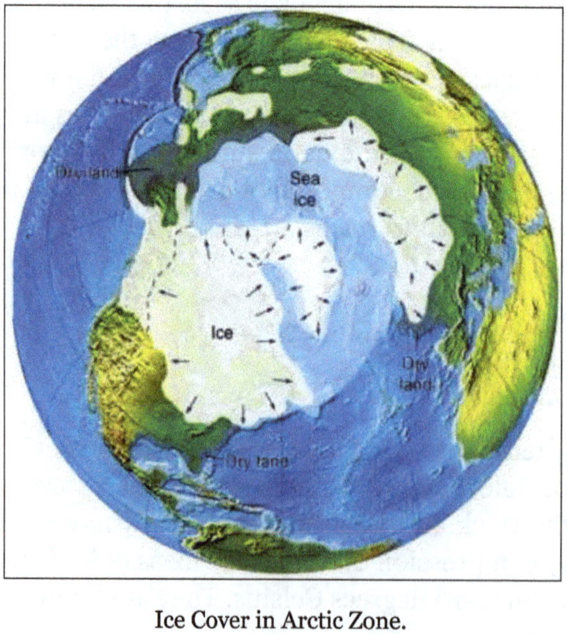

Ice Cover in Arctic Zone.

Glaciers are found on all continents except Australia. Their widespread existence is in polar areas because of favorable conditions for the accumulation of ice. Since the Arctic is an ocean around the North Pole, the amount of glacial ice is widely variable with changing seasons, but the ice survives throughout the year. The land parts in the periphery of the Arctic Ocean like northern Canada, Alaska, Greenland and northern Eurasian region have sustaining glaciers. These land masses have glaciers throughout the year. The direction of the flow of ice is outward as per the slope of the area.

Almost whole of the Antarctic continent is situated south of the Arctic Circle. Its 98 per cent area is covered with ice and only 2 per cent bare rocks are found. The Antarctica has ice sheet with an average thickness of about 2.16 Kilometers. The maximum thickness at Terre Adelie is recorded to be 4776 meter. Almost half of the coastal fringe areas is free from ice. Huge icebergs float on the ocean waters particularly in summer season (November to February) for the southern hemisphere. About 80 percent area of Greenland is under ice cover. The cross section drawn on the maps of Greenland and Antarctic has been drawn and the thickness of the ice over them is clearly seen.

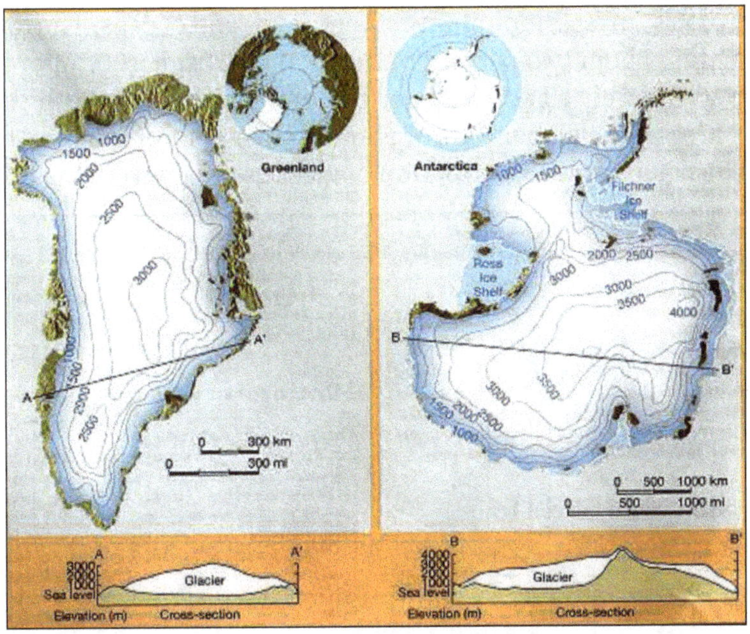

Ice Cover over Antarctic and Greenland.

We know that the effect of temperature is progressively reduced with increasing altitudes. It is the highest air temperature recorded at sea level and keeps on declining with increasing height. The distribution of temperature is also affected by the inclination of the sun rays. Due to this reason, the air temperature keeps on declining with increasing latitudes. That is why, the permanent snow line is different at different latitudes. Therefore, altitudes at various places where the ice cover is found is different with changing latitudes. In Polar Regions, it is available at the ground level. In equatorial zone, it is found five kilometers above the mean sea level whereas in temperate zone, it is one km to five kilometers. In central Africa at Kilimanjaro, it is found to be about 5500 meter, on Himalayas, it between 4500 to 6000 meter and on Alps, it is 2800 meter above mean sea level. Therefore, glaciers are found at the higher altitudes of the mountain ranges even in tropical and temperate region of the world.

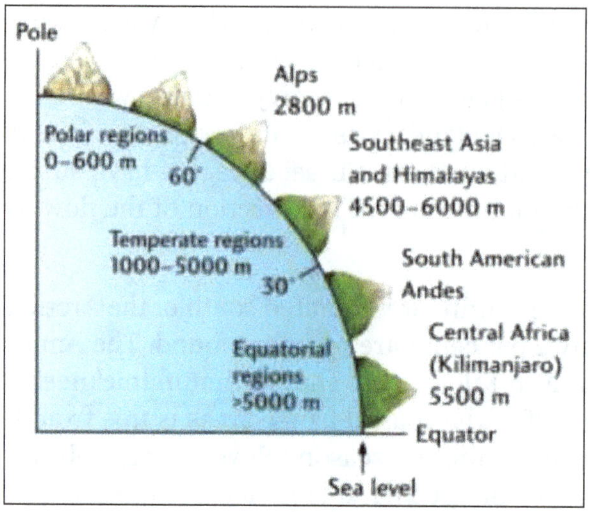

Height of Snowline at Different Latitudes.

The entire frozen water of the globe is termed as cryosphere. The cryosphere constitutes of ice and snow occupying the earth's surface, mountains, rivers, lakes and oceans. The total amount of water stored in the cryosphere is about two percent of the total hydrosphere. Out of the total area under ice is 85.30 per cent on Antarctica and 13.27 per cent in Arctic region. Both constitutes 98.57 per cent of the glaciated area of the globe. Rest of about 1.5 per cent of the glaciated area is in other parts of the world. The details of the distribution of glaciers are given in the table.

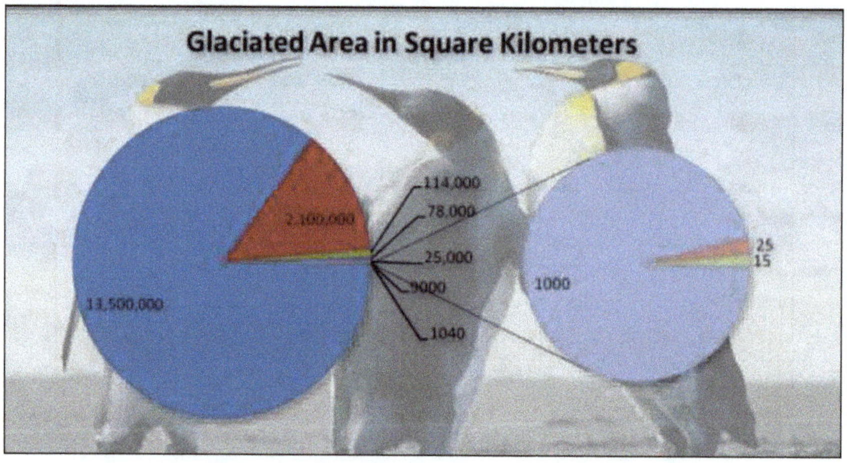

Table: Area and Approximate Volume of Present-day Glaciation.

Area/Region	Glaciated Area in km²	Glaciated Area in %	Volume of Ice km³	Volume of Ice in %
Antarctic	13,500,000	85.30	29,500,000	89.03
Arctic	2,100,000	13.27	3,200,000	9.66
Asia	114,000	0.72	220,000	0.66
North America	78,000	0.49	150,000	0.45
South America	25,000	0.16	48,000	0.14
Europe	9000	0.0569	16,000	0.05
New Zealand	1000	0.0063	1800	0.0054

Africa	25	0.0002	35	0.0001
New Guinea	15	0.00009	25	0.00008
Total	15,827,040	100.00	33,135,860	100.00

Glacier

Glacier is a creeping mass of ice. It moves very slowly. Immediate apparent movement of glacier is difficult to perceive. It takes a longer time to see the movement. The movement happens under the influence of gravity to the downslope direction. Since the glaciers are solid mass, they have enormous capacity to erode the rock and transport the rock fragments. In this process, they sculpture significant features creating distinct landscapes. But before coming to the landscapes, let us discuss about the types and their methods of erosion and transportation.

Types of Glaciers

1. Mountain Glacier: The accumulation of ice over the mountain slope covering large areas is called mountain glacier. It is the outward flow of ice from the ice-fields encompassing several peaks of the surrounding areas. It is, mostly, narrow and long because it occupies the preexisting river valley. In the higher altitude, the temperature is very low but as long as it comes down, the temperature keeps on increasing. This leads to ablation or breaking of glacier into ice pieces and ultimately melting. They are generally termed to that glaciers which are found on the mountains of tropical and temperate region. Glaciers are also found in the mountains of the polar region, but generally they are termed as polar glaciers because they occupy big continuous area of ice accumulation.

2. Valley Glacier: Valley glacier is a narrow strip of ice accumulation confined within the two walls of the preexisting river valley. Valley glaciers are also found on the mountain slopes. Sometimes, both of them are used as synonyms but they have slight difference. Mountain glaciers are huge in size whereas the valley glaciers are small in size and the ice is restricted within the two banks of the previously carved out river valleys. The downside terminating glacier is tongue shaped, i.e., smoothened round in appearance. Both of them are also known as Alpine glaciers. In fact, Alpine glaciers originate from the mountain summits and descend through the mountain valleys. Valley glaciers are very short in comparison to the continental glaciers. The longest of this type is less than 80km long. The smallest may be of a few hundred meters.

Valley Glacier

Table: Five Longest Non-Polar Glaciers in the world.

Rank	Name of the Glacier	Location	Length in km
1	Fedchenko Glacier	Tajikistan	77
2	Siachin Glacier	India	76
3	Biafo Glacier	India (POK)	67
4	Bruggen Glacier	Chile	66
5	Baltoro Glacier	India (POK)	63

3. Piedmont Glacier: When the ice mass moving through the steep surface of the mountain reaches at the foothills or the plain area, it creates piedmont glacier. Piedmont means an area from the foothills to the flat plain land. The confined ice of the valley glacier is forced to spread over a larger area because of lowering of the slope. It takes the shape of bulb-like lobes.

Piedmont Glacier

4. Continental Glacier: A continuous sheet of ice occupying a very large area is known as continental glacier. The movement of the continental glacier is, normally, from the center to outward all directions. Relatively small sized continental glaciers are known as ice fields. Similarly, a big sized continental mass of glaciers covering enormous area is called ice sheets. Almost whole of the Antarctic and Greenland are covered with ice sheets. A thick mass of ice and snow forming a glacier over a great deal of area is termed as ice cap. It may develop over the continental or on mountain glacier. Continental glaciers are very long and large in comparison to the valley glaciers. The longest continental polar glacier is more than 500 km long.

Table: Five Longest Continental Polar Glaciers in the world.

Rank	Name of the Glacier	Location	Length in km
1	Lambert-Fisher Glacier	Antarctica	77
2	Novaya Zemlya Glacier	Russia	76
3	Artic Institute Glacier	Antarctica	67
4	Nimrod-Lennox-King Glacier	Antarctica	66
5	Denman Glacier	Antarctica	63

Anatomy of Alpine/Valley Glaciers

Alpine glacier is concerned with mountain glacier found in different continents at varying altitudes. Uppermost section is characterized by zone of accumulation of ice due to extreme cold condition. Intermediaries are characterized by greater rock plucking from the glacial valley and abrasion. Steep slope is associated with crack in ice blocks called crevasses. Lowermost zone is accompanied by melting of the ice due to increase in temperature. This is known as zone of ablation. In this zone, the load carried by the glacier is deposited popularly known as moraine. Over the years, a sort of equilibrium of ice accumulation is maintained.

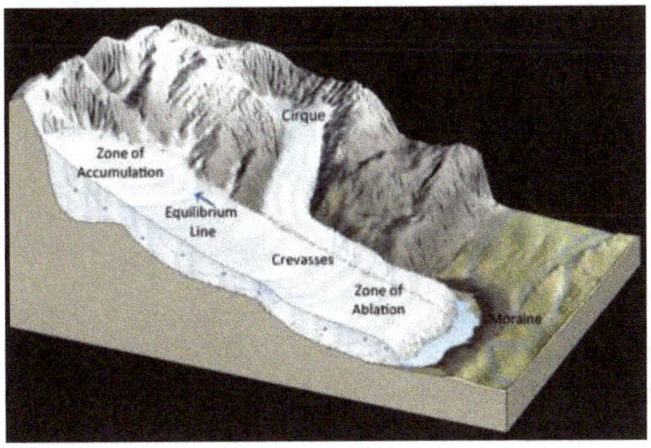

Work of Glaciers

The glaciers perform their works in three different ways – erosion, transportation and deposition.

Erosion

Glacier performs its erosional works primarily in two ways –plucking and abrasion. When any sort of irregularity is found in the way of a glacier, the solid mass of ice exerts enormous pressure. The pressure plucks and crushes the rock in the way. It becomes easier when the jointed rocks are found. This process is termed as plucking or quarrying. The plucked rocks or the fragmented rocks materials available keep on moving along with the moving ice. Once that material moves along the side or the bed of the glacier, they erode the surface. This leads to the scratching of the bed as well as the reduction of the moving materials in their size. This process is termed as abrasion. Both plucking and abrasion is very clear from the figure.

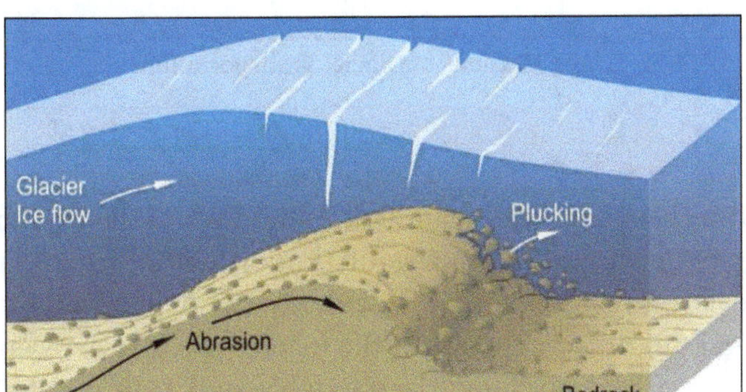

Plucking and Abrasion by Glacier.

Factors Affecting the Rate of Glacial Erosion

There are several factors affecting the glacial erosion like – thickness of ice, topography, geology of the area, temperature, velocity of the glacier, slope amount of ice accumulation and ablation etc.

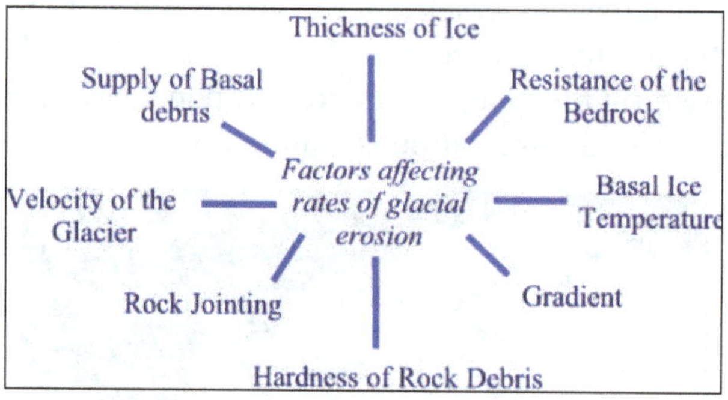

Transportation

The work of transportation by glacier is performed in three ways. When the weathered rock materials falls on the glaciers, they are transport on the ice surface. Slowly and slowly, a part of the materials fall below the ice surface and are transported for a considerable distance. Some of the rock fragments moves further downward and reaches to the bottom of the glacier. Based on the position of the rock fragments they are categorized into three. The first one is said to be surface (supraglacial), the second to be internal (englacial) and the third to be basal (subglacial). They are shown in figure.

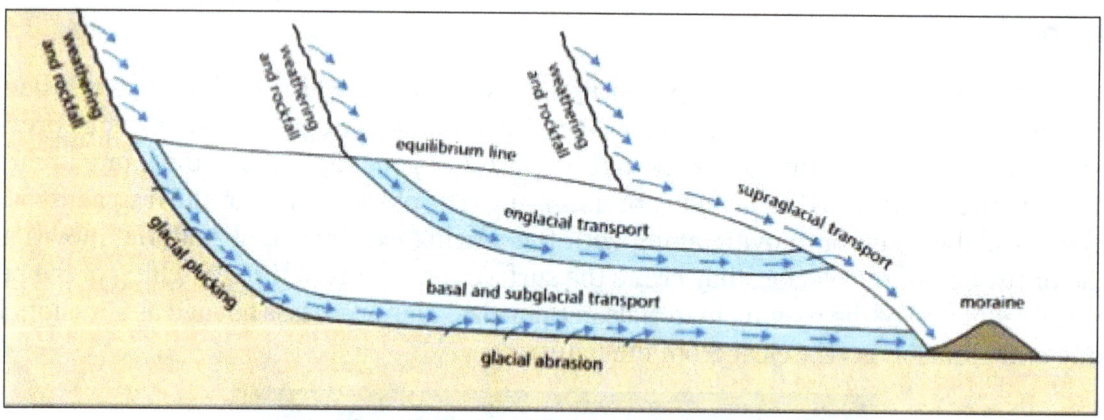

Three Methods of Glacial Transportation.

Deposition

The eroded and transported load of the glacier is deposited when the carrying capacity is reduced. The carrying capacity is reduced due to slope reduction or melt of ice and further lowering of the mass of ice. In both the cases, the deposition is evident. The deposition by glacier is very distinct because it is unsorted and ungraded deposits of the load of the glacier. Depending upon the place and material of deposits, they may be categorized in to several groups about which, it will be dealt while dealing with landforms.

Erosional Landforms

A number of landforms are created due to erosion by glaciers. Important among then are the followings:

1. Cirque: The shape of the cirque is just like an armchair and it is being made by glacial erosion. When the glacier descends from the mountain slope, it moves faster due to very steep slope. Later when the slope lowers, the excessive accumulation of ice exerts enormous pressure and the base is eroded by ice rotation, plucking and abrasion. In due course of time the base is further eroded and a depression is created. It is called cirque which is a French term and meant for a circle or bowl shaped. It is also known as Corrie in Scottish.

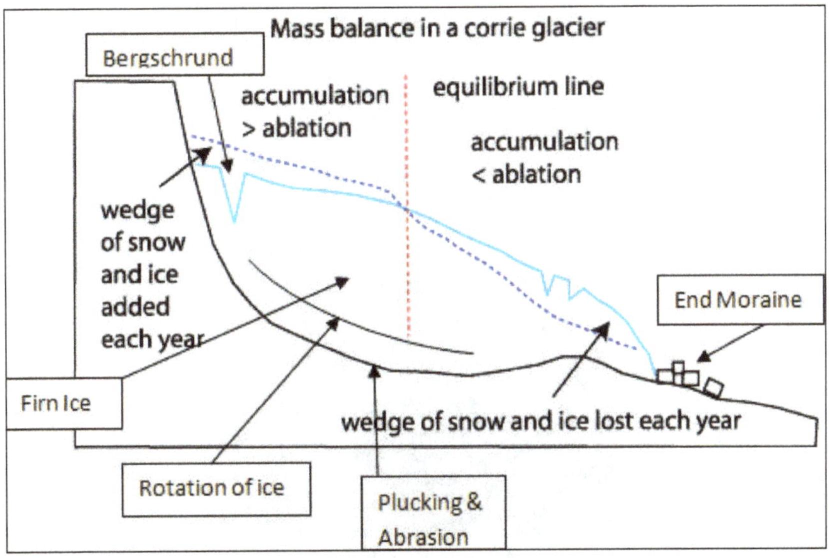

Formation of Cirque or Corrie.

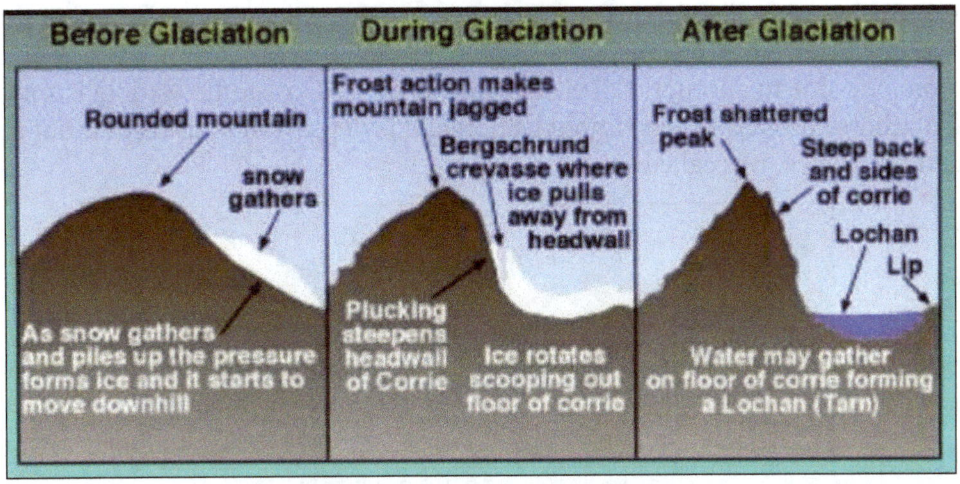

Successive Development of Cirque or Corrie.

2. Aretes: When several cirques or corries are developed on different sides of a mountain peak, their separating boundaries are further narrowed and sharpened. The sharp knife or blade edged boundaries are called arêtes. It is shown in a diagrammatic presentation in figure and the real feature in figure. Mountain climbers generally follow aretes for scaling the peak.

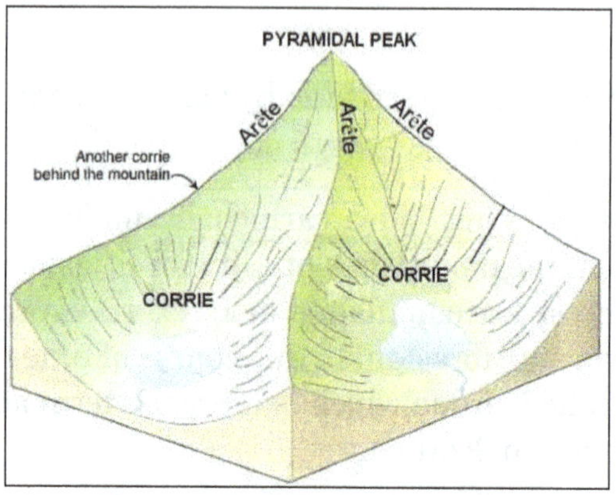

Diagrammatic Presentation of Aretes.

Arete in Real Field.

3. Col: Col is a depression like feature made between two peaks. Generally, col is made by the erosive action of glacier. This depression is created by the moving glacier. It helps in passing the mountain from one side to another. Col is also known as pass or saddle. Saddle is a seat fastened on the back of a horse for sitting and riding the horse. It is depressed at the center but raised in the front and rear and hence, it is called so.

Appearance of Col or Saddle.

4. Horn or Pyramidal Peak: When the glaciers are developed from different sides of a peak, they erode the preexisting surface of the peak successively and make it a sharp and pointed peak with very steep slope. This gives a rise of a pyramidal peak which is also termed as horn as it is very pointed.

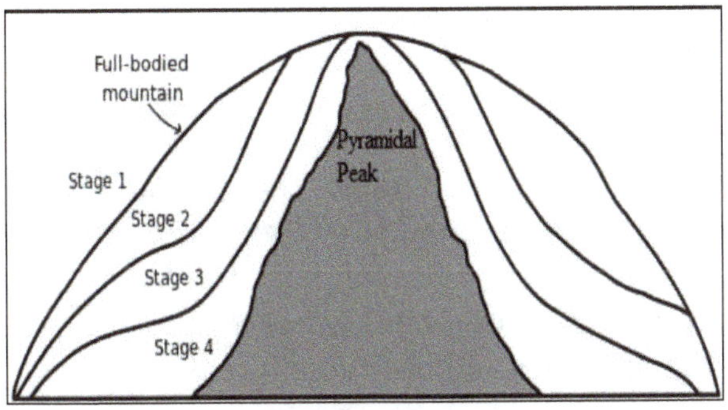

Successive Development of Pyramidal Peak.

Fully Grown Pyramidal Peak.

5. Hanging Valley: Hanging valley is a tributary glacial valley of a main glacier. Main glacier has huge amount of ice, hence, erode the base greater. Therefore, the main glacial valley is much deeper. The tributary glacier has lesser amount of ice, hence, less erosion. Therefore, the base of the tributary glacier remains above the main glacier valley. When the glaciation is over, the tributary valley is seen above the main valley. That is why, it is called hanging valley.

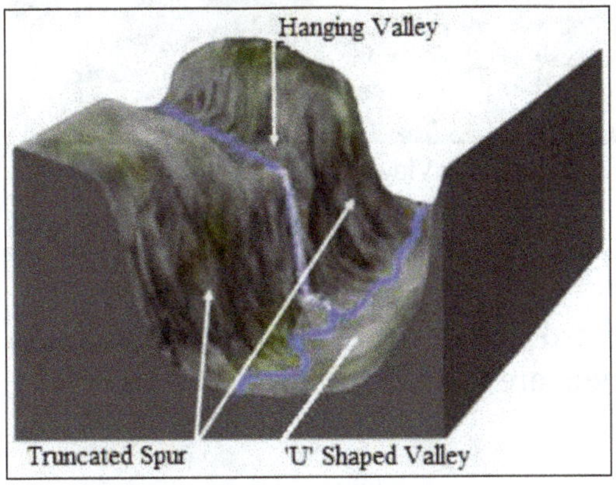

Hanging Valley and Truncated Spur.

6. Truncated Spur: Spur is a long tongue shaped ridge dropping down in an inverted 'V' separating the two adjacent valleys. The valleys confluence with another main river. When the same area comes under the influence of glacier (due to climate change), the main river turns into main glacier. Volume of ice in the main glacier is very large it erodes its valley greater. Since, the glacier is the moving solid ice it erodes not only the bed but side also. The dropped down spur is eroded and the slope of the spur is interrupted. This interrupted/ broken spur is called truncated spur.

7. 'U' Shaped Valley: Glacial ice movement erodes its sides and make the trough much wider. The bed of the glacier is flat and both sides are almost vertical. This elongated wider trough is called 'U' shaped valley. When the glaciation period is over and ice is melted, the valley is exposed.

8. Fjord: Fjords are submerged glacial valleys under the sea water along the coast. The bottom of the valley is cut down because of heavy accumulated ice movement even below the sea level. With the culmination of ice age, the sea level also rises and down cut glaciated valleys along the sea coast are drowned under the sea water. Because of this process, the coast is greatly dissected. This sort of coasts are found in the high latitudes and near polar areas.

Close-up View of Fjord along the Sea Coast in High Latitude.

Panoramic View of Fjord Coast in Norway.

Depositional Landforms

Numerous landforms are created due to deposition of glacial loads when its capacity to carry is reduced. Important among them are the followings:

Moraine

The materials carried by the glacial ice is deposited in an ideal condition when the carrying capacity is reduced. These deposits are generally unsorted and un-stratified as it is deposited as and when the material reaches and dropped. Most of them are boulders of different sizes. Depending upon the place of deposits, the moraines are classed into different groups. They are:

- Lateral Moraine: The two sides of the glacier is generally have exposed rocks. The weathered materials are dropped at the side of the glacier and are transported along. Materials are also supplied due to landslide as well. Carried materials also erode the side of the glacier and further more loads are generated. These materials are deposited along the side of the glacier and hence, they are called lateral moraine.

- Medial Moraine: When tributary glacier or two glaciers meet with each other, the meeting side's lateral moraines also joins together and moves further downward. Slowly and slowly the joined lateral moraine reaches towards the middle/ near middle. When the ice melts, the joined lateral moraine is deposited and is called medial moraine.

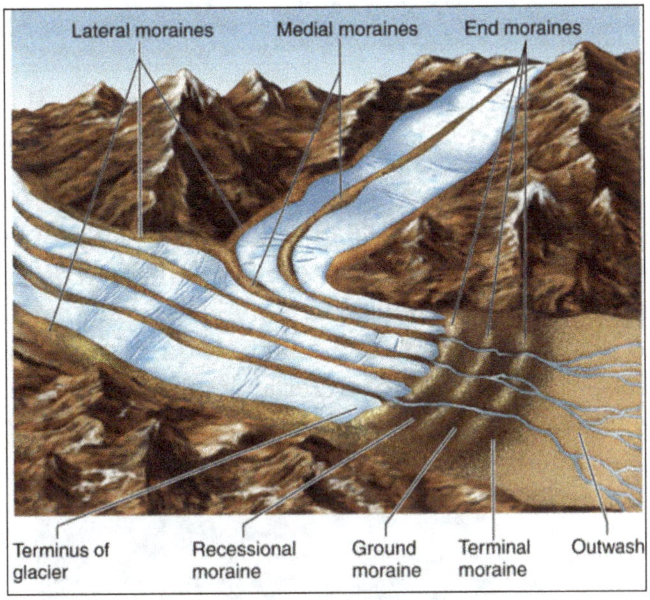

Different Types of Moraines.

- Terminal Moraine: The deposits which occur at the down end of the glacier in a ridge-like accumulation. These debris are pushed forward by advancing glacial tongue and dropped at the end. The terminal moraine spreads from one side or bank of the glacier to another covering the entire width. It appears like a belt of small hilly ground with knobs and joints.

- Recessional Moraine: Recessional moraine is a sort of terminal moraine. After the formation of terminal moraine, when the glacier retreats, a new terminal moraine is formed at upstream. The previous terminal is, later, called as recessional (down) moraine.

- Ground Moraine: Irregular blanket of glacial loads deposited on the floor in a rolling to flat landscape after glacial melt/ retreat is called ground moraine. As the term itself explains, it is the whole of the ground down side the existing glacier where the glacial load is deposited. It is formed in both the cases – continental glacier and valley glacier. Ground moraine is also known as till plain.

Glacial Till and Till Plain

The glacier carries huge amount of loads with it. When the glacier melts/ retreats, the transported loads are deposited. This deposition is undulating to flat in look. The deposition of such materials at a certain location at small scale is till and the widespread area containing till is termed as till plain. This type of till plain is very extensive in continental glacier but confined in the valley glacier.

Glacial Till

Glacial Till plain.

Outwash Plain

Down side from the glacial melt, water associated with smaller pieces of ice floats below. Gravels, sands, silts and clays are carried downward by the melted water and deposited far away. It forms a sloppy but relative flat plain known as outwash (material washed/ carried by water) plain. It is different from till plain. Till plain has unsorted materials directly deposited by melt of ice and, hence, dropped without ordering of the size of materials. Outwash plain has sorted deposits as it is deposited by glacio-fluvial (small ice pieces of ice and flowing water) action.

Outwash Plain.

Esker

Formation of Esker.

Esker is a long, narrow and zigzag ridge like structure generally made up of stratified sediments deposited by subglacial meltwater. Its height varies from 5 meter to 50 meter and width is 50 meter to 500 meter whereas the length many as 500 meter to many kilometers. Esker is not necessarily a continuous feature, but it may be a detached or broken. Literally, it is deposited by the meltwater of the glacier and hence, they are sorted.

Kettle

A kettle is a shallow and small pothole or depression area in which glacial brought sediment is deposited and remaining hollow part covers the water. It is formed and appear on the surface by retreating glacier. Probably, this depression is created by the plucking, but later deposition of sediment and water accumulation give the birth of kettle. This water-filled body is known as Kettle Lake.

Kettle and Kettle Lake.

Kame

Kame is the accumulation in a mound like shape made up of poorly sorted sand, boulders and gravels brought by glacier at its terminus. A number of kames a formed when the glacier melts and terminates or recedes. A group of closely formed kames are termed as kame complex. Several kettle lakes are seen in a kame complex.

Roche Mountonnée

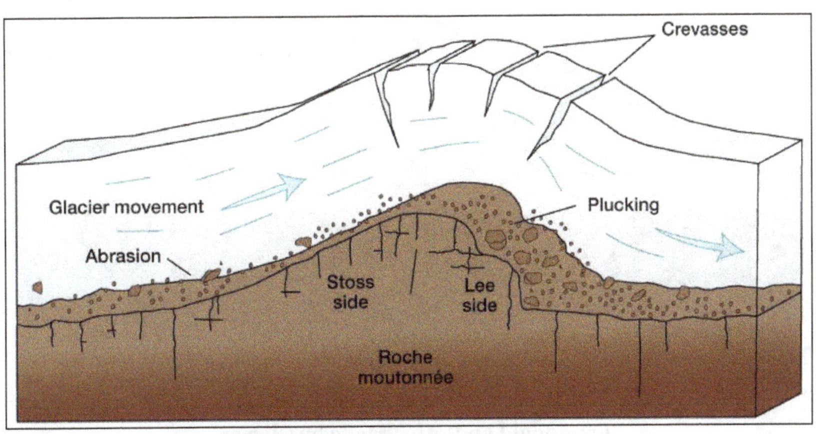

Formation of Roche Mountonnée.

It is a French word meant for woolly rock, but in English it means sheep-back or sheep-rock. Roche Moutonnée is smoothened bedrock of a glaciated region by abrasion in the upstream side and rough and steep surface to the downstream due to plucking/scouring. Once the area is free from glacier the same feature appearances to the surface.

Drumlin

Drumlin is an elongated oval-shaped deposited mound of small hill-like feature formed by glacier. It is made up of glacial till containing boulder, gravels and sand transported by glacier and finally deposited in the above mentioned shape. They are seen in the plain and are many in number. They vary in dimension like one to two kilometers in length, 400 to 600 meter in width and 15 to 30 meter in height.

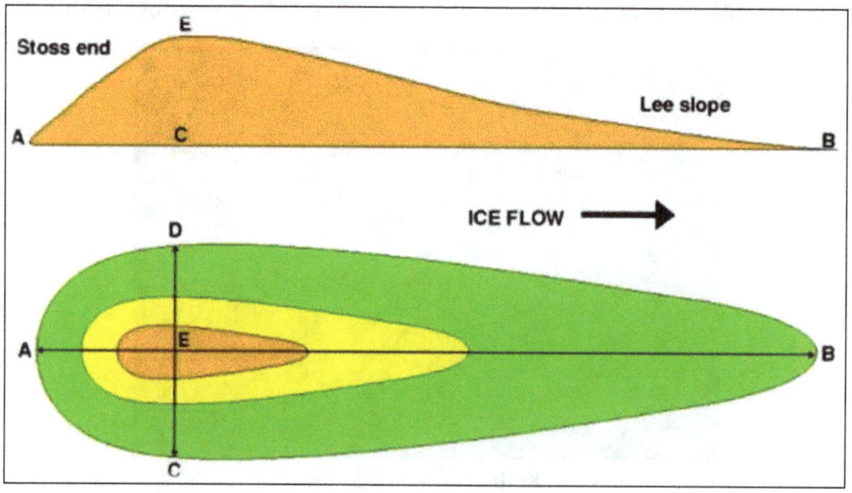

Longitudinal and Cross-Sectional View of Drumlin.

An Overview of Erosional Features

The important erosional and Depositional landforms made by glacier are represented below:

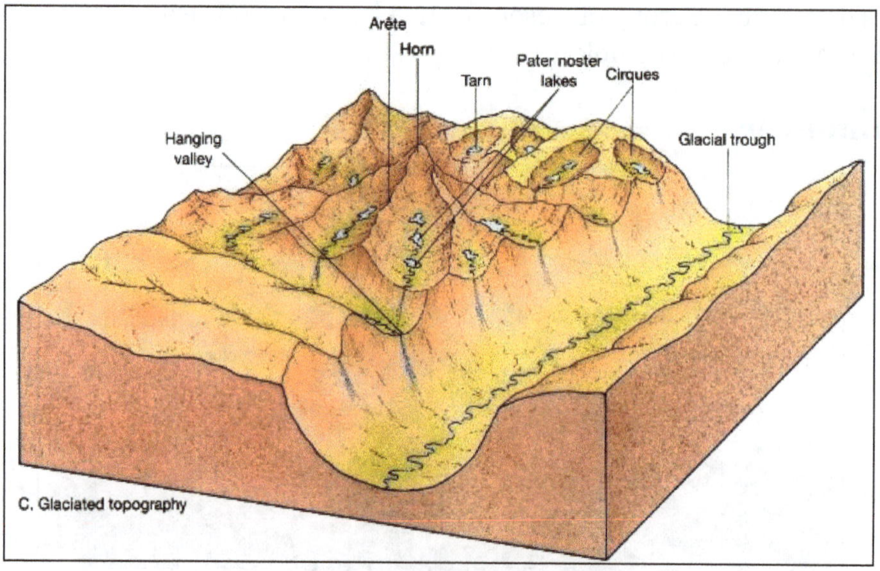

Erosional Landforms Made by Glacier.

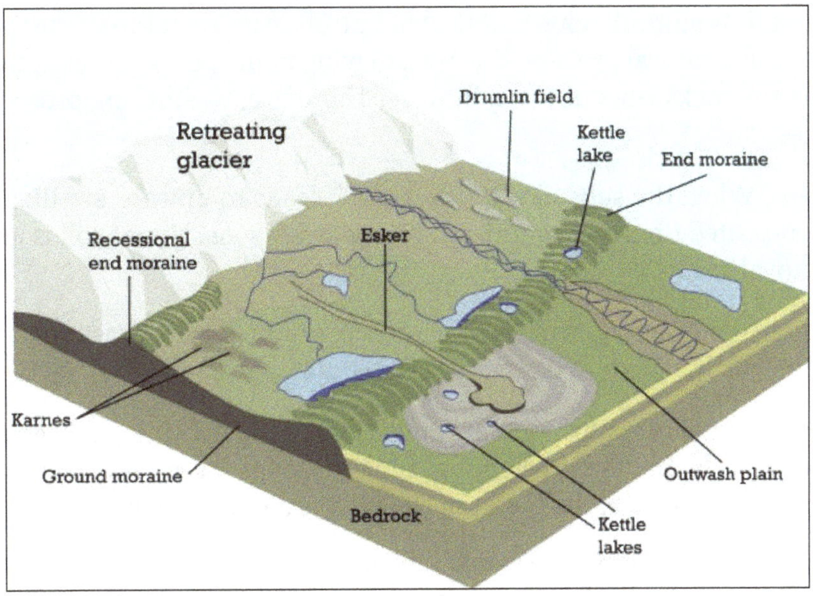

Depositional Landforms Made by Glacier.

Periglacial Landscapes

Periglacial refers to the area in the periphery of the glaciated region. During summer, the region is almost free from ice on the surface but the subsoil is still frozen throughout the year. This is very clearly be seen from Figure and thin layer of snow and ice accumulation in Figure. The seasonal freezing and thawing of subsoil is regular feature of the region. During winter, the surface witnesses heavy snowfall and ice accumulation. In another words, the area is almost permanently frozen condition but without permanent ice cover on the surface. The mean monthly temperature is between minus one degree to minus 15 degree Celsius. The periglacial areas changes with the change in the climatic condition prevailing on the planet.

Frozen Subsoil Water in Permafrost Area.

Thin Layer of Frost/Snow in Permafrost Area.

Based on the extent of freezing condition, the permafrost areas are classified into three groups – continuously permafrost, discontinuously permafrost, sporadic permafrost and isolated permafrost. The surface layer, where freezing and thawing is seen, is known as active layer. The periglacial areas witness several processes important among are frost action, frost heave, frost shattering indirectly leading to rock-fall, earthflow and debris flow.

- Frost Action: It is quite obvious that the volume of ice is greater than that of water. Freezing and thawing is a natural process. The water laying in the pores/ cracks or opening expands and breaks the rocks when it turns into ice. Therefore, weathering process is amplified by this process.
- Frost Heave: When the subsoil water freezes, it leads to upward swelling of top soil. This process is repeated time and again with changing seasons. Upper soil is loosened and subject to removal easily.
- Frost Shattering: Freezing and thawing process is known as frost shattering. The water laying in the cracks of exposed rocks in cold climates leads to breaking it.
- Rock-fall: Rock-fall is a type of mass movement in which the fragmented rocks tumble down almost freely. In the cold climates, the exposed rocks are fragmented by freezing and thawing action and hence, the credit of rock-fall is credited to permafrost action.
- Earthflow and Debris Flow: The availability of water content liquefies the soil and if it is on the slope, it flows quickly. This action is seen in the periglacial areas as the melt water of ice triggers it. Debris flow includes the unconsolidated soil, rock fragments and remnants of biotic life along with water. When these are associated with on the slope, it leads to debris flow and it is also triggered by water made available by snow melt in periglacial areas.

Landforms in Periglacial Areas

Several types of landforms are found in the periglacial areas. Some of them are:

1. Patterned Ground: Patterned ground is termed to that ground surface when the laying materials are distributed in a certain pattern. The pattern may be of any kind like stripes, ovals, circles, polygons particularly in the periglacial areas. They are made by frost heave action.

Patterned Ground.

2. Ice Wedges: The wedges mean pieces or slice or blocks. When the rocks are broken into pieces due to ice action, it is termed as ice wedges. In this situation, the water or ice penetrates vertically into surface for several meters. In cold climates (periglacial areas) the seasonal freezing and thawing cause the rock to break and penetrate further down during summer. In winter, it freezes and further the space is widened leading to rock breaking.

3. Blockfields: Blockfields are areas of huge expanse constituting the freeze-thaw weathered angular rock blocks in periglacial areas. Rock blocks get disintegrated into smaller pieces and thus, angular rocks are seen on the surface.

Angular Rocks Covered Blockfield.

4. Pingos: Pingos are hills of earth and gravel in a dome-shaped formed in the periglacial areas under which the subsoil water freezing and tends to push the upper part upward. Its height may vary from 5 meter to 70 meter and the diameter from 30 meter to one km. most of the pingos are circular in shape.

Formation of Pingos.

Hillslope Processes

Hillslopes essentially cover the whole landscape. Some are steep, others are gentle. Under the natural condition, water, sediments, and rocks move down slope. The present topography and soil thickness reflect a sensitive balance between weathering, erosion, and deposition. Human activities can upset the balance and cause:

- Loss of the fertile soil from cultivated lands.
- Excessive erosion of lands and sediment load in streams.
- Catastrophic disaster of landslides.

Planners need to understand geomorphic processes caused by the action of water on hillslope.

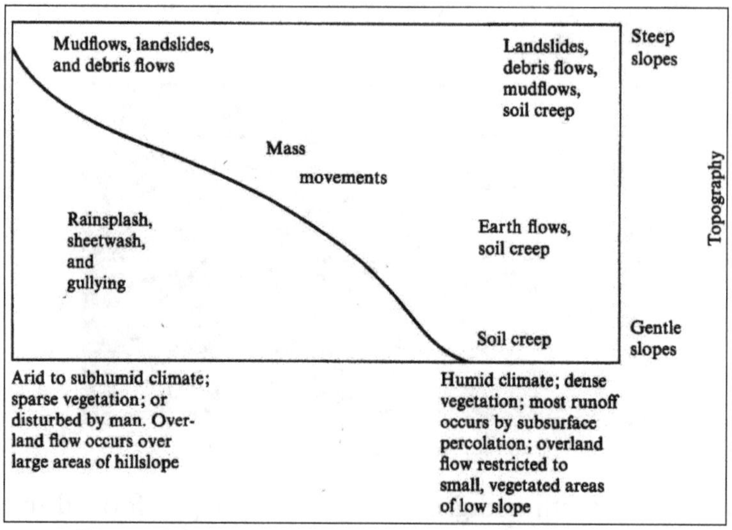

Conditions of climate, vegetation, landuse, and topography under which various geomorphic processes dominate hillslope erosion.

Normal and Accelerated Erosion

Soil profile develops over a long time. It reflects the balance of input (weathering of bed rock) and output (erosion). Normal rate of erosion keeps the soil profile stable under the current climate. Human activities can significantly accelerate erosion resulting in the soil loss.

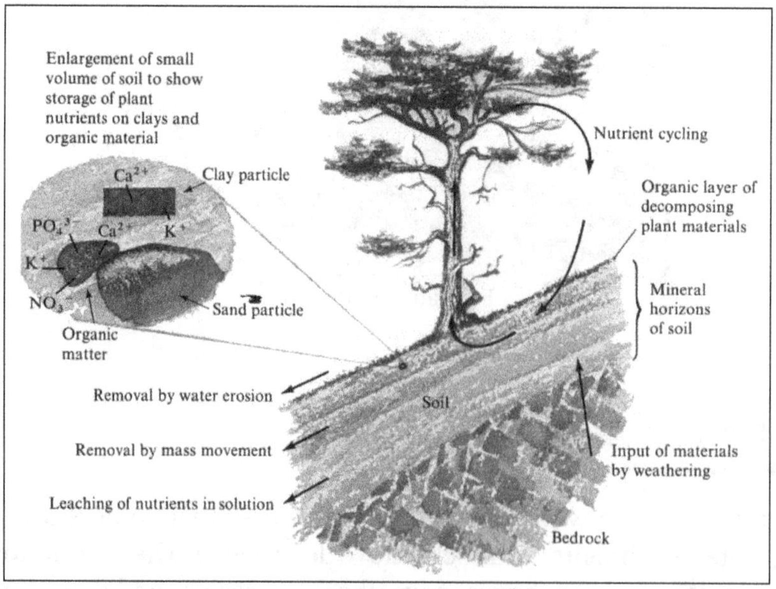

The soil achieves a balance between the input and output of materials.

Hillslope Erosion by Water

Erosion is driven by Horton overland flow, which primarily occurs on non-vegetated surface. Saturation overland flow does not contribute much because the soil surface is protected by a vegetative cover. Horton overland flow causes; rainsplash and sheetwash erosion, and gullying.

Rain Splash Erosion

A moving object posses kinetic energy (e.g. a car on a highway). This energy is released when the object hits something (highway clash). Each rain drop has kinetic energy which is released to splash soil particles upon hitting the ground. On average, the travel distance of a particle landing on the down-slope side is greater than the opposite, resulting in net downward motion of the soil. Each drop splashes a little bit, but billions of drops can cause significant erosion.

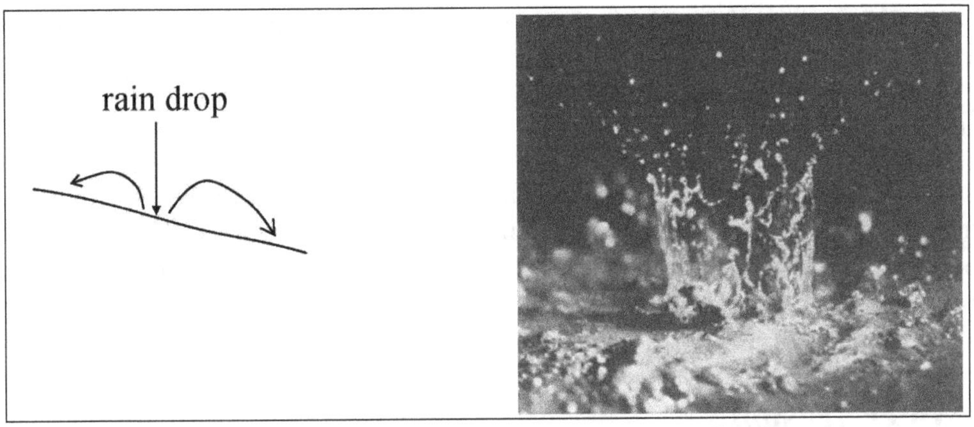

High-speed photograph of a raindrop impact on a soil surface.

This process is particularly important on steep slopes devoid of vegetation. Thick vegetation protects soils from direct hits of rain drops. Soils are resistant to erosion when they have high organic content and reasonable amount of clay to develop aggregated structure. Tillage breaks down the structure and reduce the resistance to erosion and also the infiltration capacity.

Sheetwash Erosion

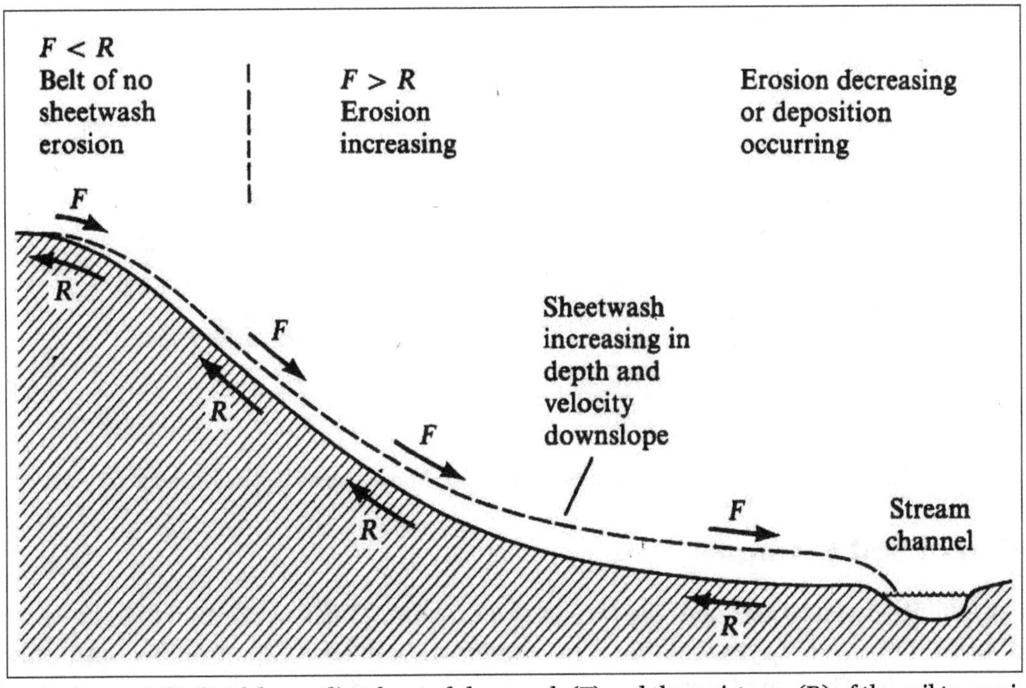

The relative magnitude of the eroding force of sheetwash (F) and the resistance (R) of the soil to erosion.

Horton overland flow results in an irregular sheet of water flowing downhill. The depth and velocity of flow increase downslope as more water is generated by precipitation excess. If the soil resistance to erosion is constant, the intensity of erosion depends on the product of water depth and hillslope gradient. In general, longer and steeper hillslopes has more intense erosion. Note that a sheet of water actually consists of many tiny streams and threads of water. We call it a sheet because, on long-term average, the erosion removes a roughly uniform depth of soil. Sheetwash erosion occurs with rainsplash erosion, and it is difficult to separate the two. Both are governed by the same factors; vegetation, slope gradient, and resistance of soil to erosion.

Rill and Gully Erosion

If the minute streams of water cut separate channels, the process is known as rill erosion. Water is concentrated in rills, and efficiency and intensity of erosion increases. When the depth of a rill exceeds about 0.3 m, it is called gully. Gully erosion produces incisions up to 10^1 m deep and 10^3 m long. Gullies occupy relatively small area. Contribution of gully erosion to overall erosion from a catchment is small (a few percent), but it has major localized effects.

Rates of Water Erosion

Thousands of experimental plots across Canada and US are used to determine the relationships between the erosion rate and various factors affecting erosion. These plots are equipped with water and sediment traps. Observations are also made in natural catchments using erosion pins.

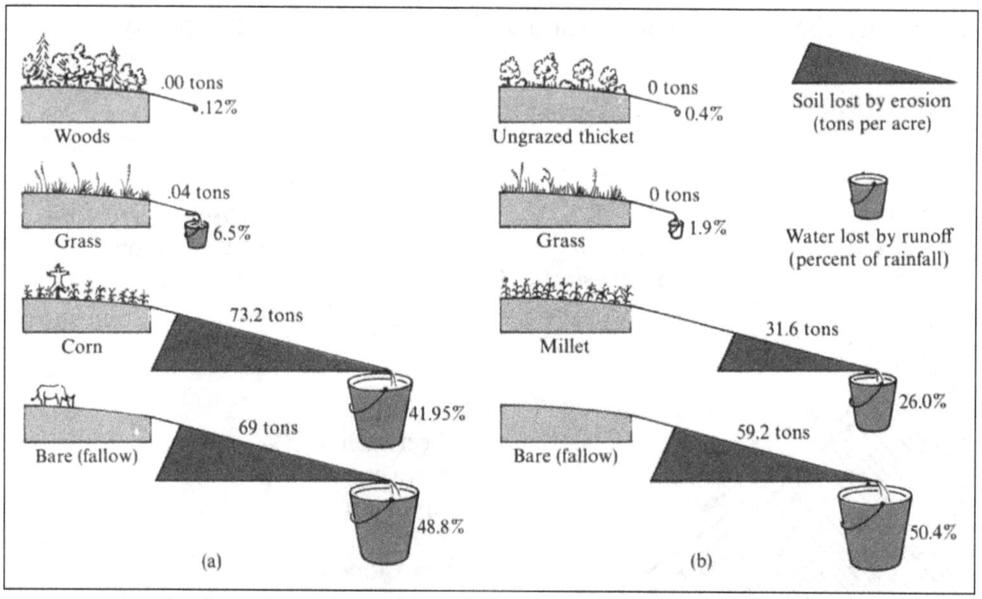

Runoff and erosion under various landuse. (a) Midwestern US. (b) Tanzania.

Prediction of Soil Erosion

Planners frequently need to estimate the rate of soil erosion. The best method is to have some local field data, but that's not always available. The Universal Soil-Loss Equation is commonly used to get a rough estimate. The use of the equation will be discussed in the lab session.

Significance of Soil Erosion

Soil erosion was recognized as a major threat to the continued productivity of the land in 1930's. Wide-spread use of fertilizers in US and Canada seemed to solve this problem, but it created other problems such as lake eutrophication. Also, some countries cannot afford applying enormous amount of fertilizers. Overall, the soil erosion must be avoided to achieve sustainable development.

Population growth in tropical and subtropical countries generates pressure on food production and energy sources (i.e. firewood and charcoal), which accelerates deforestation. Once the soil is lost, the land cannot sustain plant growth and desertification results. Soil erosion increases sediment load into streams, which endangers aquatic species and fills up reservoirs used for controlling floods and generating power.

Control of Soil Erosion and Sedimentation

- Agricultural lands:
  - Increase vegetative cover by adopting crop rotation that includes plants that provide a good cover and improve soil structure (e.g. alfalfa).
  - Avoid tilling as much as possible. Cultivate along topographic contour (contour cultivation) to increase surface detention capacity and promote infiltration.
  - Plant grains and cover crop (alfalfa) alternately as strips along topographic contour (strip cropping).
  - Cover the ground surface with hay, straw, or manure (mulching) to increase resistance and infiltration capacity.
  - Divide the hillslope into small pieces of flat lands to reduce the length and slope gradient of each piece (terracing).

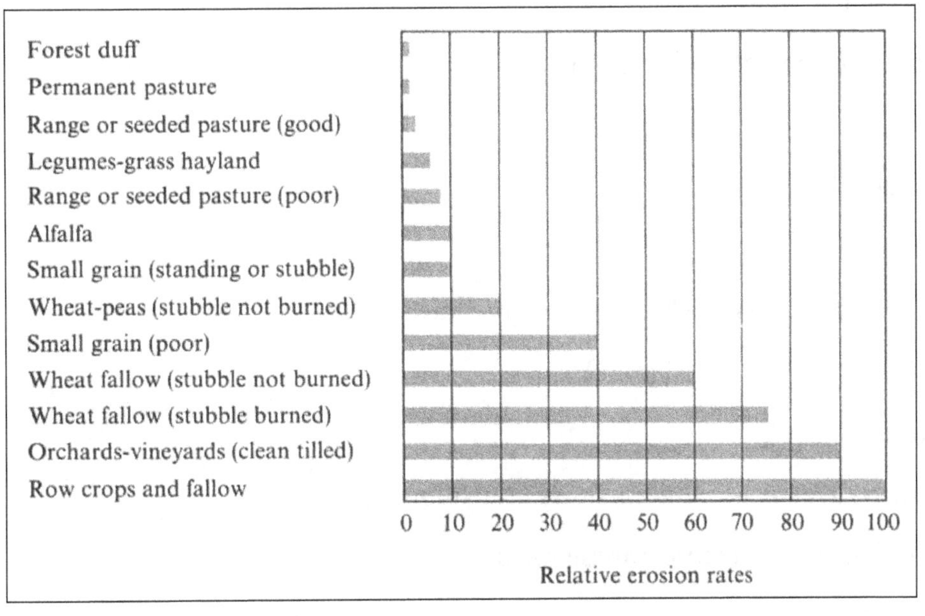

- Rangelands:
  - Control the number of stock to meet with carrying capacity of the land.
- Forest:
  - Log the trees so that forest duff layer is not destroyed.
  - Logging roads are the primary source of sediments. Try to layout logging roads on the gentlest gradient possible, preferably along ridges. Restore the area once logging operation is complete.
- Mine sites:
  - Restore the vegetation as much as possible, although this may be difficult in areas affected by acid mine drainage.
- Urbanland:
  - In construction sites, minimize the time the site is left unprotected. Maintain as much original vegetation as possible. Cover the area with surface mulch.
  - Planning is very important. Classify the lands according to soil depth, texture, and land gradient. Develop the areas based on the land classification.

Mass Wasting

This mode of hillslope processes move enormous amount of mass, up to millions of m^3, and often quite rapidly:

- Damage of houses and structures.
- Economic damage (lost farms, power lines, filled up reservoirs, etc.)
- Loss of human life.

The hillslope failures are classified as follows:

- Falls:
  - Rockfalls
- Slides:
  - Planner failures (rock slides, debris slides).
  - Rotational failures (slumps).
- Flows:
  - Debris avalanches. Debris flows. Earthflows.
  - Mudflows. Solifluction (arctic region).
- Soil creep

The classification is by no means clear-cut and combination of several processes may occur. Rockfalls occur on very steep slopes. Talus slopes and screes develop under such slopes. Cost of maintaining highways in rockfall-susceptible areas is expensive.

Table: Types of hillslope failures.

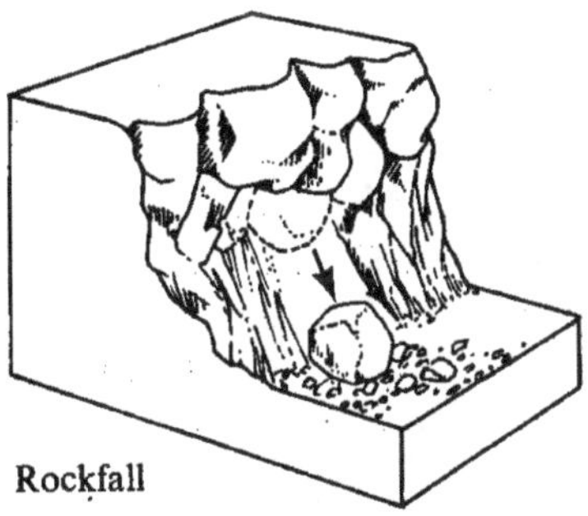

Rockfall

Extremely rapid

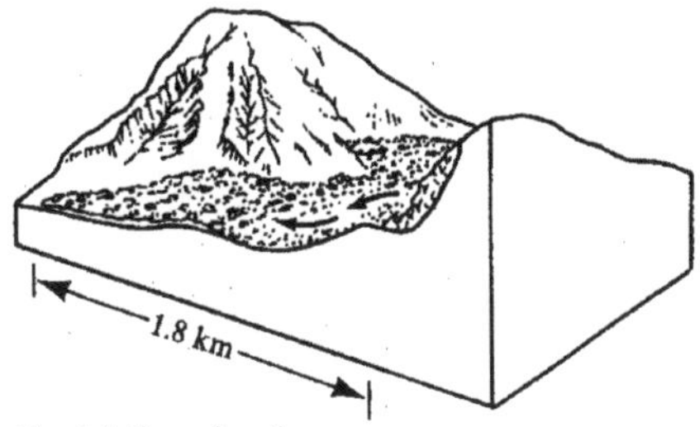

Rockfall avalanche

Extremely rapid

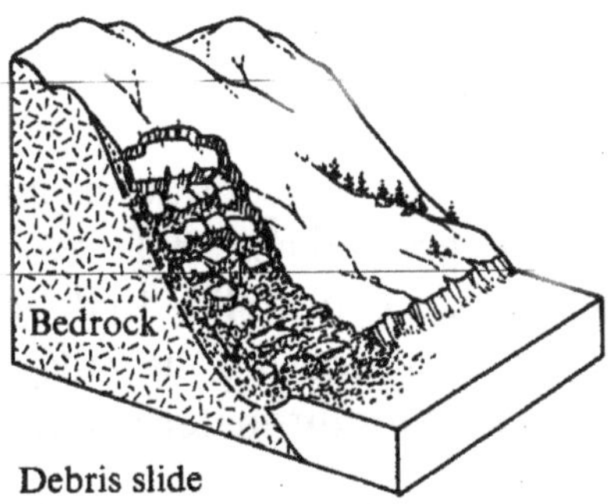

Debris slide

Very slow to rapid

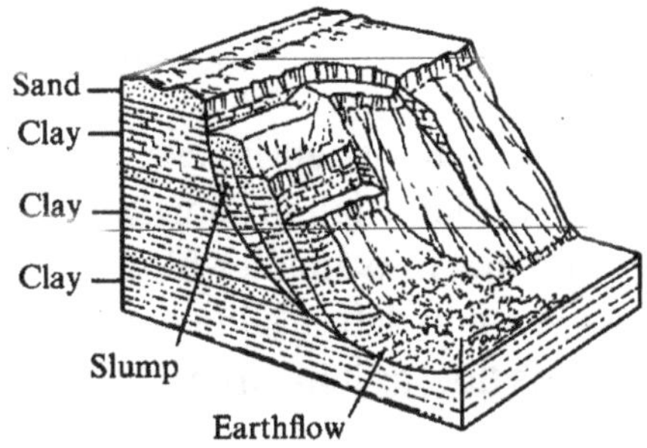

Very slow to rapid

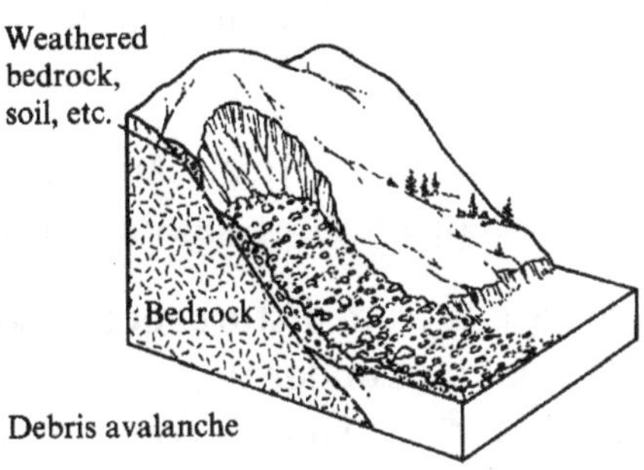

Debris avalanche

Very rapid to extremely rapid

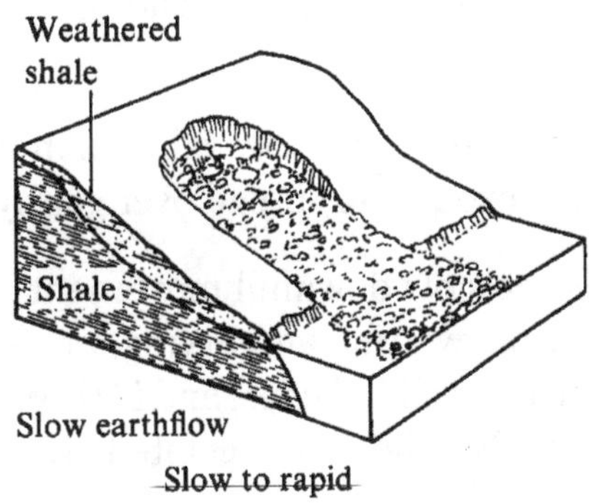

Slow earthflow

Slow to rapid

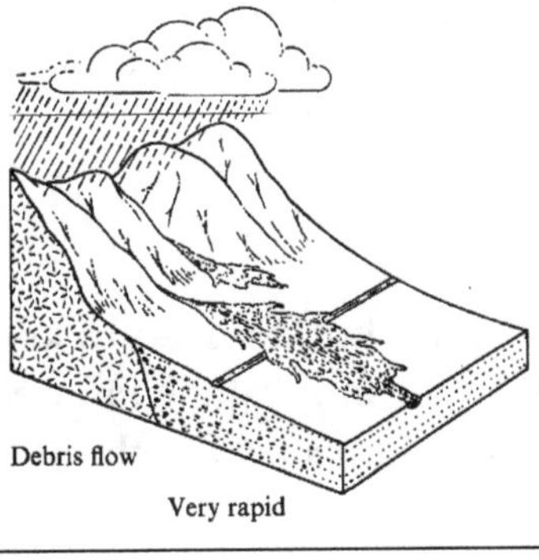

Debris flow

Very rapid

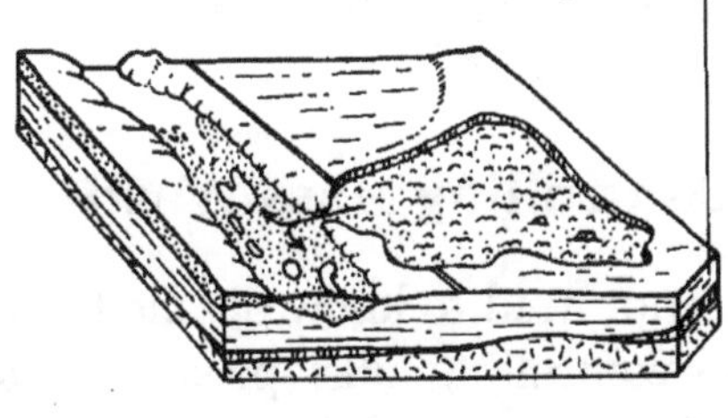

Rapid earthflow

In slides the failure occurs along a surface or within a narrow zone of deformation. The failure surface may be planer along some well-defined geological boundary, or may be arc-like. The size of relatively un-deformed blocks range from 10^0 m^3 to 10^9 m^3. A planer slide is termed rock slide or debris slide depending on the material. Where the geologic materials are deep, uniform and cohesive, rotational slumps are common. A slump block does not usually move far but causes significant ground breakage.

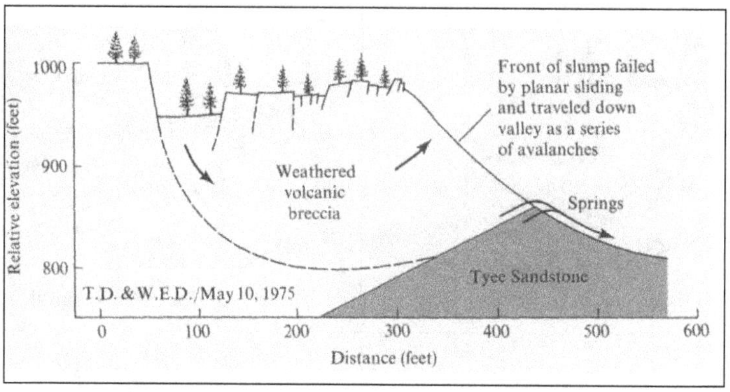

Cross section of Castle Creek Slump, Oregon.

In flows the movement of debris resembles that of a fluid; more or less continuous internal deformation of the material. High proportion of air, water and fine-grained materials favor flow. As the moving mass breaks into smaller blocks, the term changes from debris slide to debris avalanche to debris flow. A small debris slide near the head of a valley can flow down the valley and develop into a huge debris flow as it incorporate water in the valley and scour the materials along the valley.

Earthflows consist mainly of fine-grained materials. They usually occur when a slump or other failure mobilizes clays that have very high water content. Once in motion, the mass begins to flow and it develops a surging and spreading toe of dense, viscous material, which can flow over gentle gradients.

Mudflows develop where mass failures in fine-grained material mix with streams. The mudflows travel down canyons and spread over gently sloping fans of debris at the mountain front.

Soil creep is the slow downhill movement of debris that results from disturbance of soil by freezing and thawing, wetting and drying, or slow plastic deformation under the soil's own weight. It occurs on virtually all hillsides, and is responsible for most of the downslope transport of debris to streams channels in heavily vegetated areas.

Solifluction is the slow, viscous downslope flow of waterlogged soil underlain by an impervious layer, most commonly frozen ground. It is gaining considerable importance in development of arctic and subarctic regions.

Factors Controlling Mass Wasting

Planners do not need to know detailed engineering mechanics, but do need to know the nature of the forces controlling hillslope stability and the role of water in hillslope failures. In this simple case of the soil block sliding along the bedrock surface W indicates weight of the block. The downslope component $W\sin\alpha$ is the driving force.

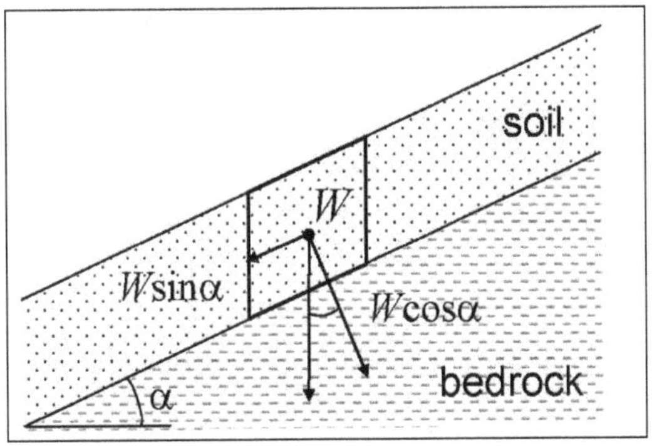

If Wsinα exceeds shear strength of the material, the slope failure may occur. The shear strength S is given by:

$$S = C_1 + C_2(W\cos\alpha - P)$$

where C_1 and C_2 are constants and P is the pressure of water.

Noting that sinα increases and cosα decreases as α increases, failure is more likely in steeper hillslopes. Also note that S decreases as P increases. Large pressure of pore water weakens the material. This is why groundwater is so important in hillslope processes. Suppose the water table rises during a storm. The pressure P at point A increases suddenly. This reduces the shear strength and causes a rotational failure.

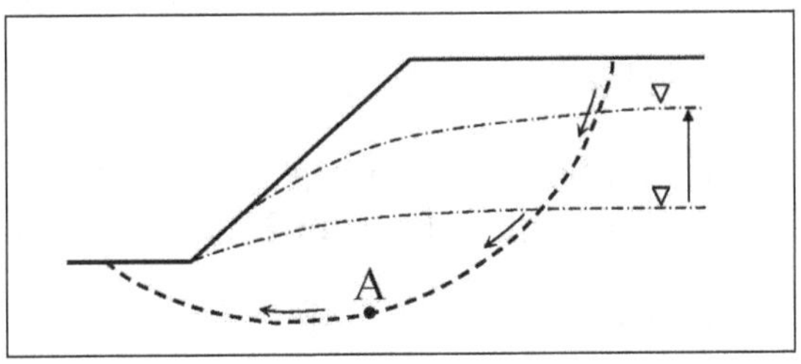

Geological Factors

Earth quakes can cause short-period increases in the downslope force and trigger landslides. In tectonically active areas, slow tilting of the ground by mountain building forces tend to increase slope gradients. The nature keeps adjusting gradients by occasional slope failures.

Human Factors

Undercutting during construction of highways and railroads increases the average slope gradients, and also increases the chance of slope failure.

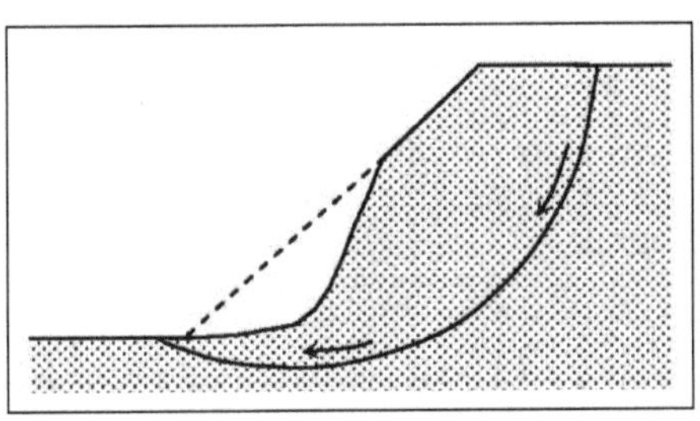

Overloading the top of hillslopes by housing construction is common. This extra weight may increase the chance of slope failure. Altering the hydrology may have dramatic effects on hillslope stability. For example, clear cutting trees promote soil erosion and weaken the support of soils by tree roots. It also reduces evapotranspiration and raise the water table.

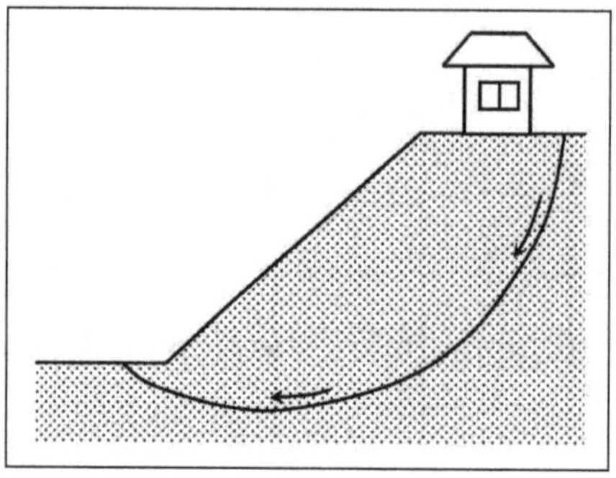

Recognition and Avoidance of Landslide Hazards

Detailed analysis of hillslope stability requires the expertise of engineers. Planners need to decide:

- Whether a site is in a stable area.
- Whether there is enough uncertainty to warrant a detailed site investigation by an expert.
- Whether the site is so obviously unstable that it should be avoided.

Techniques for evaluating landslide hazard involve recognizing at least one of the following:

- Past hillslope failures: Direct evidence in airphoto, indirect evidence of altered vegetation, subtle topographic features, deposits formed by hillslope failures.
- Conditions that is conductive to hillslope failures: Steep slope gradients, mechanically weak geological material, poor permeability, high water table, seepage in the vicinity of steep slope.
- De-stabilizing effects of planned development: Undercutting, overloading, changing hydrologic conditions.

References

- Aeolian-features-and-processes-3905272: geoscienceworld.org, Retrieved 09, May 2020
- Fluvial-processes-and-landforms-309630899: researchgate.net, Retrieved 22, February 2020
- Marine-processes: alevelgeography.com, Retrieved 17, July 2020

Chapter 4

Types of Weathering

The process of breaking down of rocks and minerals on the surface of the Earth by water, atmospheric gases and biological organisms is termed as weathering. It can be categorized into physical, chemical and biological weathering. The topics elaborated in this chapter will help in gaining a better perspective about weathering.

Weathering is the first stage in the denudation (wearing away) of the exposed landscape. It is a process by which surface and subsurface rocks break up, dissolve, and decompose. All weathering activity occurs on the spot (in situ). Since Weathering disintegrates, dissolves, decomposes the rock, it facilities the movement of rock debris by mass wasting. It is a part of exogenic process because it directly and indirectly derives required energy from insolation, which guides systems above the surface of the earth, including the atmosphere. Weathering is also an important process of earth system because it facilitates in the formation of mineral and soil development. 60 per cent population of our country is directly and indirectly dependent on agriculture which is to a great deal dependent on soil.

Weathering is the alteration and breakdown of rock minerals and rock masses when they are exposed to the atmosphere. Weathering processes occur in situ, that is, in the same place, with no major movement of rock materials involved. Weathering is a fundamental Earth process. Weathering changes rocks from a hard state, to become much softer and weaker, making them more easily eroded.

Weathering Grades

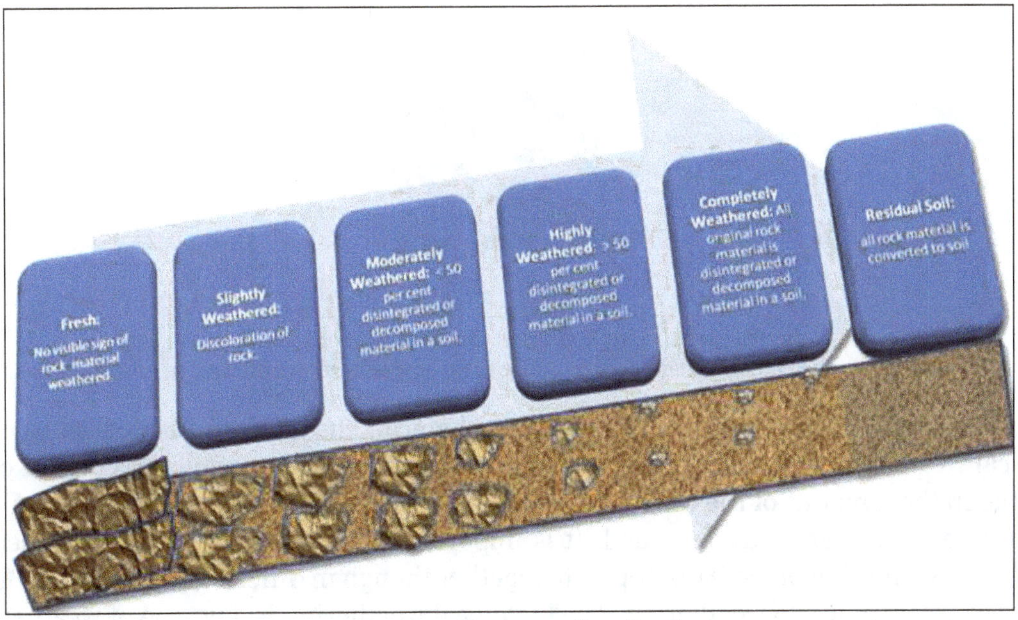

Scale of weathering grades of rock.

We can also identify the different weathering grades of rock.. Stage wise there are six grades of weathering. "Weathering grades of rock" ranges between fresh weathering grade to residual soil. The fresh grade exhibits no visible sign of rock material. The slightly weathered stage is identified by discoloration of rock. When less than 50 per cent rock material is disintegrated or decomposed in a soil it is called moderately weathered stage in relation to present grade. In the highly weathered category more than 50 per cent rock material is disintegrated or decomposed in a soil. In the last and final stage all the rock material is converted to soil which is known as residual soil.

It is important to note that the rate of weathering depends on property of parent rock (mineral solubility and rock structure), climatic conditions (frequency of frost, rainfall, temperature and length of exposure to sunlight). The presence of organic activity in the soil also influences the rate of weathering. Weathering processes are largely determined by the climate and vegetation of a place. Dry locations are largely influenced by physical weathering and moist places by chemical weathering.

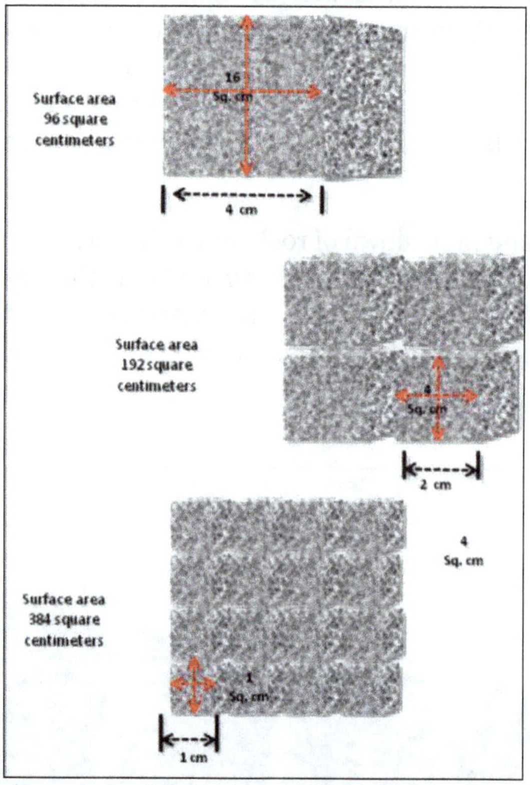

The physical weathering generates smaller pieces or rock which in turn produces additional surface area for chemical weathering.

In general there are three types of weathering: mechanical, chemical and biological weathering. Mechanical weathering denotes the breakdown of rock without any conspicuous degree of chemical change in the minerals of the rock. Chemical weathering involves the decomposition of rocks in which decay of minerals are recorded. It is important to note that in real world mechanical and chemical weathering process may operate together though in differing proportions. Biological processes also contribute to weathering. The biological weathering is any type of weathering that is caused by living organisms.

Physical Weathering

Physical weathering is the disintegration of rock and soil aggregates, by physical (mechanical) processes acting primarily on pre-existing fractures (e.g. joints, cracks between mineral grains); reduces size of fragments according to rock and soil structure (producing grains, crystals, blocks, slabs, etc.), with no change in composition.

Processes

- Stress (pressure) release: Disintegration of rock in parallel sheets as it expands in response to the removal of confining stress:
    - Most common mechanism of stress release is removal of overlying rock by erosion; thus this process is controlled by erosion but subsequently controls erosion.
    - The dilation fractures conform to the surface topography and increase in spacing with depth (e.g. From a few cm at the surface to a few metres at 30 m in the sierra nevada mountains of yosemite national park).
    - Thermal contraction due to cooling counteracts expansion; therefore stress release is most pronounced near the surface where the rocks have already cooled and contracted.
    - Also most common in massive rocks: Higher thermal conductivity causes heat loss and thus reduces counter influence of cooling and contraction, fractured or thinly bedded rock will expand without sheeting, massive rocks store stress until overburden pressure is very low (i.e. About 100 m of overburden).
    - Stress release causes exfoliation: The separation of concentric layers of rock.
- Thermal expansion and contraction (insolation weathering):
    - The surface temperature of dark colored rock can vary from 0-50°C between day and night, since rock (esp. Jointed rock) has low thermal conductivity.
    - The differential stresses of expansion and contraction of the outer 1-5 cm of rock causes separation of concentric shallow layers (another form of exfoliation) called spalling or spheroidal weathering when it effects boulders.
    - Controversy about effectiveness:
        - Re: The ability of solar radiation to generate sufficient heating and cooling.
        - Rocks disintegrate after fires, especially rocks composed of minerals with varying coefficients of volumetric expansion (e.g. Granite: volume of quarts increases 3x more than that of feldspar; versus greater resistance of fine-grained rocks).
        - Dry granite heated and cooled from 30 to 140°C for 89,400 cycles over 3 years (equivalent of 244 years of diurnal cycles) produced no perceptible change, even with microscopic examination.

        - But, 244 years is small amount of geologic time.

- Growth of foreign crystals (salt weathering):
    - Mainly hydrated salts which are water soluble at normal ranges of atmospheric temperature and humidity; they hydrate and dehydrate repeatedly generating considerable stresses in fractures and between grain boundaries in permeable rock.
    - Mostly granular disintegration.
    - Minerals are transported in solution and precipitate as soil and groundwater evaporate; thus most effective in desert landscapes where water tables are near the surface.
    - Origin of salt: Sea water, chemical weathering of marine or evaporite sediments, dissolved in snow and rain, precipitates in lakes.
    - E.g. Gypsum: Relatively insoluble except in acid rain (dilute sulfuric acid), thus enhanced weathering with acid rain.
- Hydration (slaking):
    - Wetting, swelling and disintegration of soil aggregates, layered and fine grained rocks.
    - Also pressure of air drawn into pores under dry conditions and then trapped as water advances into soil and rock; suction or -ve pore pressure (less than atmospheric) can exert considerable stress.
    - E.g. Biotite expands 40% by volume contributing to the weathering of granite.

Expansion of Clay Minerals by Volume	
Ca-montmorillonite	45-185%
Na-montmorillonite (bentonite)	1400-1600%
illite	15-120%
kaolinite	5-60%

- Frost shattering and hydration shattering:
    - Freezing of water in pores and fractures.
    - The specific volume (vol./unit mass) of water increases by 9% upon freezing producing stress that is greater than the tensile strength of all common rocks.
    - Therefore the stress generated by the crystallization of ice is the most pervasive mechanism of weathering, effects all rocks.
    - However, the effectiveness of freezing water is influenced by:
        - Lack of confinement: If more than 20% of pore space is empty, then the tensile stress may be less than the tensile strength, thus frost shattering is most effective in saturated rock; ice under pressure deforms plastically and thus will extrude through daylighted fractures and pores.

- Decreased freezing point with increasing pressure (supercooled water) and impurities (e.g. Salt).
- Necessary frequency and magnitude: Extent of frost shattering is a function of the combination of frequency, duration and intensity (rapidity and degree) of freeze-thaw cycles.

- Hypotheses to account for apparent effectiveness of frost shattering; hydrofracturing:
    - Thin (monomolecular) films of water do not freeze even at low temperatures, given strong capillary adhesion to rock; powerful molecular forces in these thin films of semicrystalline water draw water along microfractures opening and propagating them.
    - Shallow freezing forces films of capillary water along microfractures, disintegrating rocks well below the depth of freezing; e.g. hydrofracturing in the Allegheny Mountains (southern Appalachians of West Virginia and Pennsylvania) extends to 12-15 m, while frozen ground rarely extends below 1 m.

- Plants:
    - Minor agent of weathering.
    - Maintain cracks created by other processes.
    - Roots may pry rocks apart when tall trees sway in a strong wind; root throw can break fragments away from bedrock.
    - As lichens expand and contract or are removed by abrasion they can pull small rock fragments loose.

Chemical Weathering

Since many rocks and minerals are formed under conditions present deep within the Earth, when they arrive near the surface as a result of uplift and erosion, they encounter conditions very different from those under which they originally formed. Among the conditions present near the Earth's surface that are different from those deep within the Earth are:

- Lower Temperature (Near the surface T = 0-50°C).
- Lower Pressure (Near the surface P = 1 to several hundred atmospheres).
- Higher free water (there is a lot of liquid water near the surface, compared with deep in the Earth).
- Higher free oxygen (although O_2 is the most abundant element in the crust, most of it is tied up bonded into silicate and oxide minerals - at the surface there is much more free oxygen, particularly in the atmosphere).

Because of these differing conditions, minerals in rocks react with their new environment to produce new minerals that are stable under conditions near the surface. Minerals that are stable under P, T, H_2O, and O_2 conditions near the surface are, in order of most stable to least stable:

- Iron oxides, Aluminum oxides - such as hematite Fe_2O_3, & gibbsite $Al(OH)_3$.
- Quartz
- Clay Minerals
- Muscovite
- Alkali Feldspar
- Biotite
- Amphiboles
- Pyroxenes
- Ca-rich plagioclase
- Olivine

These are igneous minerals that crystallize from a liquid. Note the minerals that occur low on this list are the minerals that crystallize at high temperature from magma. The higher the temperature of crystallization, the less stable are these minerals at the low temperature found near the Earth's surface. The main agent responsible for chemical weathering reactions is water and weak acids formed in water.

- An acid is solution that has abundant free H^+ ions.
- The most common weak acid that occurs in surface waters is carbonic acid.
- Carbonic acid is produced in rainwater by reaction of the water with carbon dioxide (CO_2) gas in the atmosphere.

$$\underset{\text{water}}{H_2O} + \underset{\text{carbon dioxide}}{CO_2} \rightarrow \underset{\text{carbonic acid}}{H_2CO_3} \rightarrow \underset{\text{hydrogen ion}}{H^+} + \underset{\text{bicarbonate ion}}{HCO_3^-}$$

H^+ is a small ion and can easily enter crystal structures, releasing other ions into the water.

Types of Chemical Weathering Reactions

- Hydrolysis: H^+ or OH^- replaces an ion in the mineral. Example:

$$\underset{\text{Orthoclase}}{4KAlSi_3O_8} + \underset{\text{Hydrogen ion}}{4H^+} + \underset{\text{Water}}{2H_2O} \rightarrow \underset{\text{Potassium ion}}{4K^+} + \underset{\text{Kaolinite(clay mineral)}}{Al_4Si_4O_{10}(OH)_8} + \underset{\text{quartz}}{8SiO_2}$$

- Leaching: Ions are removed by dissolution into water. In the example above we say that the K^+ ion was leached.
- Oxidation: Since free oxygen (O_2) is more common near the Earth's surface, it may react

with minerals to change the oxidation state of an ion. This is more common in Fe (iron) bearing minerals, since Fe can have several oxidation states, Fe, Fe^{+2}, Fe^{+3}. Deep in the Earth the most common oxidation state of Fe is Fe^{+2}.

$$\underset{\text{Pyroxene}}{3Fe^{+2}SiO_3} + \underset{\text{Oxygen}}{1/2O_2} \rightarrow \underset{\text{Magnetite}}{Fe_3O_4} + \underset{\text{Quartz}}{3SiO_2}$$

- Dehydration: Removal of H_2O or OH^- ion from a mineral.

$$\underset{\text{Goethite}}{2FeO \cdot OH} \rightarrow \underset{\text{Hematite}}{Fe_2O_3} + \underset{\text{water}}{H_2O}$$

- Complete Dissolution: All of the mineral is completely dissolved by the water.

$$\underset{\text{Calcite}}{CaCO_3} + \underset{\text{Carbonic Acid}}{H_2CO_3} \rightarrow \underset{\text{Calcium ion}}{Ca^{2+}} + \underset{\text{Bicarbonate ion}}{2(HCO_3)^-}$$

- Living Organisms: Organisms like plants, fungi, lichen, and bacteria can secrete organic acids that can cause dissolution of minerals to extract nutrients. The role of microorganisms like bacteria has only recent been discovered.

Weathering of Common Rocks

Rock	Primary Minerals	Residual Minerals*	Leached Ions
Granite	Feldspars	Clay Minerals	Na^+, K^+
	Micas	Clay Minerals	K^+
	Quartz	Quartz	---
	Fe-Mg Minerals	Clay Minerals + Hematite + Goethite	Mg^{+2}
Basalt	Feldspars	Clay Minerals	Na^+, Ca^{+2}
	Fe-Mg Minerals	Clay Minerals	Mg^{+2}
	Magnetite	Hematite, Goethite	---
Limestone	Calcite	None	Ca^{+2}, CO_3^{-2}
*Residual Minerals = Minerals stable at the Earth's surface and left in the rock after weathering.			

Interaction of Physical and Chemical Weathering

Since chemical weathering occurs on the surface of minerals, the water and acids that control chemical weathering require access to the surface. Physical weathering breaks the rock to provide that surface. Fracturing the rocks, as occurs during jointing, increases the surface area that can be exposed to weathering and also provides pathways for water to enter the rock. As chemical weathering proceeds, new softer minerals, like oxides or clay minerals, will create zones of weakness in rock that will allow for further physical weathering. Dissolution of minerals will remove material that holds the rock together, thus making it weaker.

When rock weathers, it usually does so by working inward from a surface that is exposed to the weathering process. If joints and fractures in rock beneath the surface form a 3-dimensional network, the rock will be broken into cube like pieces separated by the fractures. Water can penetrate more easily along these fractures, and each of the cube-like pieces will begin to weather inward.

The rate of weathering will be greatest along the corners of each cube, followed by the edges, and finally the faces of the cubes. As a result the cube will weather into a spherical shape, with un-weathered rock in the center and weathered rock toward the outside. Such progression of weathering is referred to as spheroidal weathering.

Factors that Influence Weathering

- Rock Type and Structure:
    - Different rocks are composed of different minerals, and each mineral has a different susceptibility to weathering. For example a granite consisting mostly of quartz is already composed of a mineral that is very stable on the Earth's surface, and will not weather much in comparison to limestone, composed entirely of calcite, which will eventually dissolve completely in a wet climate.
    - Bedding planes, joints, and fractures, all provide pathways for the entry of water. A rock with lots of these features will weather more rapidly than a massive rock containing no bedding planes, joints, or fractures.
    - If there are large contrasts in the susceptibility to weathering within a large body of rock, the more susceptible parts of the rock will weather faster than the more resistant portions of the rock. This will result in differential weathering.

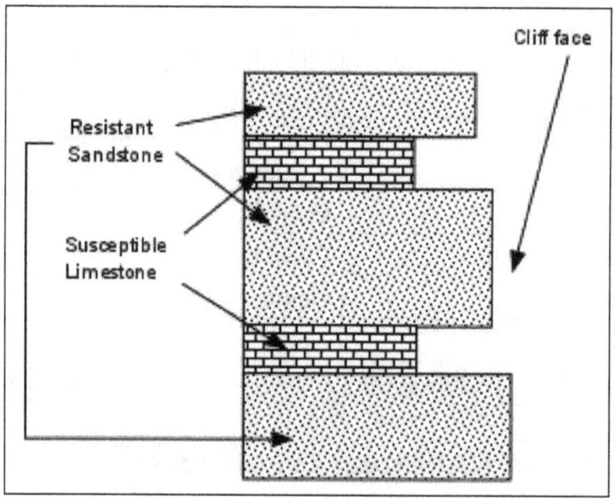

- Slope: On steep slopes weathering products may be quickly washed away by rains. On gentle slopes the weathering products accumulate. On gentle slopes water may stay in contact with rock for longer periods of time, and thus result in higher weathering rates.
- Climate: High amounts of water and higher temperatures generally cause chemical reactions to run faster. Thus warm humid climates generally have more highly weathered rock, and rates of weathering are higher than in cold dry climates. Example: limestones in a dry desert climate are very resistant to weathering, but limestones in a tropical climate weather very rapidly.
- Animals: Burrowing organisms like rodents, earthworms, & ants, bring material to the surface were it can be exposed to the agents of weathering.

Biological Weathering

- Biological weathering only refers to weathering caused by plants, animals, fungi, and microorganisms such as bacteria.
- It is contributed to or removal of ions and minerals from the weathering environment and physical variations due to movement or development of organisms.
- It is also the wearying and subsequent fragmentation of rock by plants, animals, and microbes.

Agents of Biological Weathering

Biological Weathering by Microorganisms:

- Wedging and burrowing by organisms like termites, earthworms, rodents, etc. help in showing the new surfaces to chemical attack and helps in the penetration of air and moisture.
- Bacteria, mosses, algae, and lichens frequently grow on rock surfaces, particularly in humid areas.
- They form weak acids, which can convert some of the minerals to clay.
- Algae growth can deteriorate several rock types and make it more exposed to weathering.

Biological Weathering by Humans:

- Humans also play an important role in biological weathering.
- Construction activities like road building, mining also causes weathering.
- Human beings by disturbing vegetation, ploughing and cultivating soils, also help in blending and producing new contacts between water, minerals, and air in the earth materials.

Biological Weathering by Plants and Animals:

- Decomposition of plant and animal help in the creation of carbonic acids, humic and other acids which boost decay and solubility of some elements.
- Roots of plants exert tremendous pressure on the earth materials mechanically breaking them apart.

References

- Weathering: tulane.edu, Retrieved 12, March 2020
- Ncert-notes-biological-weathering: byjus.com, Retrieved 28, June 2020

Chapter 5

Landforms in Geomorphology

The features created by the erosive and constructive actions of the wind are known as aeolian landforms whereas the landforms produced by the action of glaciers are termed as glacial landforms. This chapter closely examines the key concepts of aeolian, glacial and fluvial processes and landforms to provide an extensive understanding of the subject.

Landforms are the natural features and shapes existent on the face of the earth. Landforms possess many different physical characteristics and are spread out throughout the planet. Together, landforms constitute a specific terrain and their physical arrangement in the landscape forms what is termed as topography. The physical features of landforms include slope, elevation, rock exposure, stratification and rock type.

Oceans and continents illustrate the largest grouping of landforms. They are they further subcategorized into many different landforms based on their physical features and shapes. Examples of distinctive landforms include mountains, valleys, plateaus, glaciers, hills, loess, deserts, shorelines, and plains. Features such as volcanoes, lakes, rivers, mid-ocean ridges, and the great ocean basins are also part of landform features.

Different Major Landforms on Earth

Major types of landforms on earth include mountains, valleys, plateaus, glaciers, hills, loess, plains and desserts.

Mountains

Mountains are lands physical features protruding high beyond the hills and very high up the land surface with steep top commonly shaped up to a peak. They are created through the action of incredible forces in the earth such as volcanic eruptions. Often, mountains occur in the ocean compared to land and some are seen as mountain islands as their peaks protrude out of the water. Mountain formation result from the forces of erosion, volcanism, or uplifts in the earth's crust.

The forces of heat and pressure within the earth's interior are the main influencing factors to these forces as stated by geologists. These forces can be summed up as the plate tectonic movements – theoretically defined as the division of the earth's outmost layer into several plates which are in constant motion. Hence, the uplifts are cause by collision or pulling apart of the plates that also triggers other various geologic activities such as the ejection of magma onto the surface or volcanic eruptions.

The movements also contribute to horizontal compression that is the deformation of crustal strata which gives rise to folds. The Himalayas and the Europe's Jura and Alps mountains are examples of mountains formed as a result of horizontal compression. Some mountain ranges are also formed as a result of wind, water, and ice erosion. Other mountains are created from volcanism.

Examples of volcanic mountains include Mount Fuji in Japan, Mount Vesuvius in Italy, Mount Erebus in Antarctica, and Mount Saint Helens in the United States. Majority of volcanic mountains have summit craters that still expel debris and steam.

Valleys

A valley is a lowland area or surface depression of the earth between higher lands such as mountains or hills. In simple terms, it can be defined as a natural trough bounded by mountains or hills on the surface of the earth sloping down to the lake, ocean or stream, which is created because of water or ice erosion. On this basis, the rivers or streams flowing through the valley empty the land's precipitation into the oceans.

The lowest parts of the valleys are very fertile and make very good farmlands. Majority of the valleys on land are made up of running streams and rivers and nearly all their floors slope downstream. Valleys within the mountains normally have narrow floors. The sides of a valley are termed as valley slopes or valley walls and the section of floor along riverbanks are referred to as flood plains.

Valleys physical features include U-shaped and V-shaped caused through the forces of erosion by the flowing masses that persistently widens and deepens the valley. The flowing masses are either water or glacier that carries away huge amounts of debris. Very narrow and deep valleys are known as canyons.

Plateaus

Plateaus are fairly flat areas higher than the land surrounding it. The surrounding areas may have very steep slopes. Some plateaus such as the Tibet are situated between mountain ranges. Plateaus cover wide land areas and together with their enclosed basins they cover approximately 45% of the entire earth's land surface.

Some plateaus, for instance the Columbia Plateau of the United States and the Deccan of India are basaltic and were created because of lava flows spreading to thousands of square kilometers thereby building up the fairly flat land surfaces. Other plateaus form as a result of upward folding while some are due to the erosion of the nearby land that leaves them elevated. Because plateaus are elevated, they are subject to erosion.

Low plateaus make up good farming regions whereas high plateaus are considered great for grazing livestock. Most of the world's high plateaus are deserts. Other typical examples of plateaus include the Bolivian plateau in South America, the Colorado plateau of the United States, the Laurentian Plateau and the plateaus of Iran, Arabia, and Anatolia.

Glaciers

Glaciers are the perennial ice sheets on the planet. They are huge masses of ice that slowly move over the land surface, predominant in high mountains and the cold Polar Regions. The very low temperatures in the regions are the enabling factor for the buildup of snow and densification into ice at depths of 15 meters or even more. Most glaciers have density thickness in the ranges of 91 to 3000 meters.

The movements begin when the compaction is so dense that it moves under the pressure of its weight. It is estimated that more than 75% of the world's fresh water is currently locked away in these frozen reservoirs. The glaciers include the Greenland Ice Sheet and the Antarctic Ice Sheet. The Antarctic Ice sheets outlet glaciers comprise the steep and extensively long and narrow depression Beordmore Glacier, which is one of the longest outlets in the world. The gradual rice in continental temperatures has seen the glacial density grow smaller owing to melting.

Hills

Hills are raised areas on the surface of the earth with distinctive summits, but are not as high as mountains. Hills are created as a result of accumulation of rock debris or sand deposited by wind and glaciers. They can also be created by faulting when the faults go slightly upwards. Hills are generally present in low mountain valleys and plains.

The Black Hills are the most known. Deep erosions of areas previously raised by the earth's crust disturbances carry most of the soil away leaving behind a hill. Human activities may also create hill when soils are dug and piled giant masses. Volcanic eruptions as well create hills after the eruption when the molten materials or lava cools and hardens in a pile.

Loess

Loess is a fine-grained unstratified accumulation of clay and silt deposited by the wind. It appears brown or yellowish in color and is brought about by past glacial activity in an area. In precise, it is sedimentary deposits of clay and silt mineral particles which take place on land in some parts of the world. The thickness of loess deposits are just a few meters and one of their basic feature is known as the 'cat steps'.

It's held together by few clay particles and is mostly composed of quartz crystals which readily slide against each other. This property makes it highly susceptible to erosion which leads to the 'cat steps' feature. Loess formed after the ice age when the glaciers covering a relatively large portion of the earth melted and was carried away, exposing the vast plains of mud.

Upon drying of the mud, the forces of wind blew away the mud and exposed sediments and eventually deposited them as silt in stacks on top of each other to create bold steep banks. Regions made of loess are witnessed in eastern China and the northwestern region of the United States.

Plains

Plains are broad flat areas on the earth's surface stretching over a wide area. Plains are lower than the land in their surrounding and can be found both inland and along the coast. Coastal plains rise from the seal level up to the point they meet raised landforms such as plateaus or mountains. The Atlantic Coastal plain is a prime example of a substantially populated and fertile coastal plain.

On the other hand, inland plains are generally found at high altitudes. Thick forests normally flourish on plains in humid climates. A fairly large portion of plains are covered by grasslands, for instance, the Great Plains in the United States. Human populations prefer settling on plains because of the soil and the terrain which is good for farming and building settlements such as cities, residential areas, and transportation networks. Flood plains are also in this category and they are formed as a result of continuous accumulation of sand, silt, and mud when rivers overflow its banks.

Deserts

Deserts are the hot and dry areas of the world. They are the arid and semi-arid lands with little or no vegetation. Deserts constitute approximately 20% of the earth's total land cover and are distinguished by little or no rainfall. The deserts are divided into four major categories including the Semi-Arid Deserts, the Hot and Dry Deserts, the Cold Deserts, and the Coastal Deserts.

These deserts are located in different areas of the world. Deserts experience very high temperatures, less cloud cover, low humidity, low atmospheric pressure, and very little rain, which makes them have very little vegetation cover. The soil cover is also rocky and shallow and with very little organic matter and as such, it only supports a few plants adapted to the conditions.

Plants such as cacti and short shrubs are the ones adapted to the desert conditions because they can conserve water and tolerate the high temperatures. Animals in the deserts include insects, small carnivores, snakes, lizards, and birds adapted to survive with very little water. These animals hide during the day till nightfall to avoid the heat. An example of a desert is the Sahara of North Africa.

Aeolian Landforms

Aeolian landforms are landforms created by the erosive and constructive effect of wind on the Earth's surface. The landforms are not only limited to the earth but have also been observed on other planets such as Mars where wind action is present. Wind transports and deposits sediments of various sizes in the areas where it is the chief agent of erosion. Such particles include silt, clay, and sand among others. Winds are mostly effective in areas where there is sparse vegetation, little or no soil moisture and unconsolidated elements. Such conditions are widely present in arid environments such as deserts. The Wind erodes the earth surface by both abrasion and deflation. Deflation refers to the removal of the loose particles on the surface of the Earth by the impact of wind. Abrasion is the wearing down of the surface of particles by the grinding action of the materials carried by the wind. The material eroded by winds are transported through saltation, suspension, and skipping on the earth's surface. The mode of transportation is largely dependent on their sizes and the strength of the wind. The deposition process by wind holds clues as to the past and present directions and intensities of wind. Wind-deposited bodies occur as sand sheets, ripples, and dunes.

Dune

In physical geography, a dune is a hill of sand formed by eolian (aeolian, or wind-related) processes. Dunes can take different forms and sizes, based on their interaction with the wind. Most kinds of dunes are longer on the windward side, where the sand is pushed up the dune, and are shorter on the "slip face" on the lee side (the side sheltered from the wind). The "valley," or trough, between dunes is called a slack. A dune field is an area covered by extensive sand dunes. Large dune fields are known as ergs.

Erg Awbari (Idehan Ubari) in the Sahara desert region of Fezzan in Libya.

Some coastal areas have one or more sets of dunes running parallel to the shoreline directly inland from the beach. In most cases, the dunes serve to protect the land against potential ravages by storm waves from the sea. Although the most widely distributed dunes are those associated with coastal regions, the largest complexes of dunes are found inland in dry regions and associated with ancient lake beds or sea beds.

Mesquite Flat Dunes in Death Valley National Park.

Dunes can also be formed under the action of water flow (alluvial processes), on sand or gravel beds of rivers, estuaries, and the seabed.

One of the main problems posed by sand dunes is their encroachment on human habitats. Sand dunes move by different means, all of them aided by the wind. Sand dunes threaten buildings and crops in Africa, the Middle East, and China. Preventing sand dunes from overwhelming cities and agricultural areas has become a priority for the United Nations Environment Program. On the other hand, dune habitats provide niches for highly specialized plants and animals, including numerous rare and endangered species. Some countries have developed extensive programs for dune protection.

Dune Shapes

Crescentic

Crescent-shaped mounds are generally wider than they are long. The slipface is on the dune's concave side. These dunes form under winds that blow from one direction, and they also are known as barchans, or transverse dunes. Some types of crescentic dunes move faster over desert surfaces than any other type of dune. A group of dunes moved more than 100 meters (m) per year between 1954 and 1959 in the People's Republic of China's Ningxia Province; similar rates have been recorded in the Western Desert of Egypt. The largest crescentic dunes on Earth, with mean crest-to-crest widths of more than 3 kilometers (km), are present in in China's Taklamakan Desert.

A diagram showing the formation of a dune with a slipface.

Linear

Straight or slightly sinuous sand ridges typically much longer than they are wide are known as linear dunes. They may be more than 160 kilometers long. Linear dunes may occur as isolated ridges, but they generally form sets of parallel ridges separated by miles of sand, gravel or rocky interdune corridors. Some linear dunes merge to form Y-shaped compound dunes. Many form in bidirectional wind regimes. The long axes of these dunes extend in the resultant direction of sand movement.

Linear loess hills known as pahas appear superficially similar. These hills appear to have been formed during the last ice age under permafrost conditions dominated by sparse tundra vegetation.

Star

Radially symmetrical, star dunes are pyramidal sand mounds with slipfaces on three or more arms that radiate from the high center of the mound. They tend to accumulate in areas with multidirectional wind regimes. Star dunes grow upward rather than laterally. They dominate the Grand Erg Oriental of the Sahara. In other deserts, they occur around the margins of the sand seas, particularly near topographic barriers. In the southeast Badain Jaran Desert of China, the star dunes are up to 500 meters tall and may be the tallest dunes on Earth.

Dome

Oval or circular mounds that generally lack a slipface, dome dunes are rare and occur at the far upwind margins of sand seas.

Parabolic

U-shaped mounds of sand with convex noses trailed by elongated arms are parabolic dunes. Sometimes these dunes are called U-shaped, blowout, or hairpin dunes, and they are well known in coastal deserts. Unlike crescent shaped dunes, their crests point upwind. The elongated arms of parabolic dunes follow rather than lead because they have been fixed by vegetation, while the bulk of the sand in the dune migrates forward.

Longitudinal (Seif) and Transverse Dunes

Longitudinal dunes (also called Seif dunes, after the Arabic word for "sword"), elongate parallel to the prevailing wind, possibly caused by a larger dune having its smaller sides blown away. Seif dunes are sharp-crested and are common in the Sahara. They range up to 300 m (900 ft) in height and 300 km (200 mi) in length. In the southern third of the Arabian Peninsula, a region called the Empty Quarter because of its total lack of population, a vast erg called Rub al Khali contains seif dunes that stretch for almost 200 km and reach heights of over 300 m.

Seif dunes.

Seif dunes are thought to develop from barchans if a change of wind direction occurs. The new wind direction will lead to the development of a new wing and the over development of one of the original wings. If the prevailing wind then becomes dominant for a lengthy period of time, the dune will revert to its barchan form, with one exaggerated wing. Should the strong wind then return the exaggerated wing will further extend so that eventually it will be supplied with sand when the prevailing wind returns. The wing will continue to grow under both wind conditions, thus producing a seif dune. On a seif dune, the slip face develops on the side facing away from the strong wind, while the slip face of a barchan faces the direction of movement. In the sheltered troughs between highly developed seif dunes, barchans may be formed because the wind is unidirectional.

A transverse dune is perpendicular to the prevailing wind, probably caused by a steady buildup of sand on an already existing minuscule mound.

Reversing Dunes

Occurring wherever winds periodically reverse direction, reversing dunes are varieties of any of the above shapes. These dunes typically have major and minor slipfaces oriented in opposite directions.

Complex dune: Dune 7 in the Namib desert, one of the tallest in the world.

All these dune shapes may occur in three forms: Simple, compound, and complex. Simple dunes are basic forms with a minimum number of slipfaces that define the geometric type. Compound dunes are large dunes on which smaller dunes of similar type and slipface orientation are superimposed, and complex dunes are combinations of two or more dune types. A crescentic dune with a star dune superimposed on its crest is the most common complex dune. Simple dunes represent a wind regime that has not changed in intensity or direction since the formation of the dune, while compound and complex dunes suggest that the intensity and direction of the wind has changed.

Dune Types

Sub-aqueous Dunes

Sub-aqueous (underwater) dunes form on a bed of sand or gravel under the actions of water flow. They are ubiquitous in natural channels such as rivers and estuaries, and also form in engineered canals and pipelines. Dunes move downstream as the upstream slope is eroded and the sediment deposited on the downstream or lee slope.

The Great Dune of Pyla is the largest dune in Europe.

These dunes most often form as a continuous "train" of dunes, showing remarkable similarity in wavelength and height.

Dunes on the bed of a channel significantly increase flow resistance, their presence and growth playing a major part in river flooding.

Lithified Dunes

A lithified (consolidated) sand dune is a type of sandstone that is formed when a marine or aeolian sand dune becomes compacted and hardened. Once in this form, water passing through the rock can carry and deposit minerals, which can alter the hue of the rock. Cross-bedded layers of stacks of lithified dunes can produce the cross-hatching patterns, such as those seen in Zion National Park.

A local slang term used for these consolidated dunes is "slickrock," a name used by pioneers of the old west because their steel-rimmed wagon wheels found it difficult to gain traction on this rock.

Coastal Dunes

Dunes form where "constructive waves" encourage the accumulation of sand, and where prevailing onshore winds blow this sand inland. There is a need for obstacles (for example, vegetation, pebbles) to trap the moving sand grains. As the sand grains get trapped they start to accumulate, this is the start of dune formation. The wind then starts to affect the mound of sand by eroding sand particles from the windward side and depositing them on the leeward side. Gradually this action causes the dune to "migrate" inland; as it does so it accumulates more and more sand. These dunes provide shelter from the wind.

Coastal dunes on the Kurnell Peninsula.

Ecological Succession on Coastal Dunes

As a dune forms, plant succession occurs. The conditions on an embryo dune are harsh, with salt spray from the sea carried on strong winds. The dune is well drained and often dry, and composed of calcium carbonate from seashells. Rotting seaweed, brought in by storm waves adds nutrients to allow pioneer species to colonize the dune. These pioneer species are marram grass, sea wort grass, and other sea grasses in the UK. These plants are well adapted to the harsh conditions of the foredune, typically having deep roots which reach the water table, root nodules that produce nitrogen compounds, and protected stoma, reducing transpiration. Also, the deep roots bind the sand together, and the dune grows into a fore dune as more sand is blown over the grasses. The grasses add nitrogen to the soil, meaning other, less hardy plants can then colonize the dunes. Typically these are heathers and gorses. These too are adapted to the low soil water content and have small, prickly leaves which reduce transpiration. Heathers add humus to the soil, but have a pH of lower than 7, so make the soil slightly acidic. Heathers are usually replaced by coniferous trees which can tolerate the low pH. Coniferous forests and heathland are common climax communities for sand dune systems.

Coastal dunes in Curonian spit.

Young dunes are called yellow dunes, dunes which have high humus content are called gray dunes. Leaching occurs on the dunes, washing humus into the slacks, and the slacks may be much more developed than the exposed tops of the dunes. It is usually in the slacks that more rare species are developed and there is a tendency for the dune slacks soil to be waterlogged and where only marsh plants can survive. These plants would include creeping willow, cotton grass, reeds, and rushes. Also, there is a tendency for natterjack toads to breed there.

Desertification

One of the biggest problems posed by sand dunes is their encroachment on human habitats. Sand dunes move through a few different means, all of them helped along by the wind. One way that dunes can move is through saltation, where sand particles skip along the ground like a rock thrown across a pond might skip across the water's surface. When these skipping particles land, they may knock into other particles and cause them to skip as well. With slightly stronger winds, particles collide in mid-air, causing sheet flows. In a major dust storm, dunes may move tens of meters through such sheet flows. Like snow, sand avalanches can fall down the steep slopes of the dunes facing away from the winds, also moving the dunes forward.

Sand threatens buildings and crops in Africa, the Middle East and China. Drenching sand dunes with oil stops their migration, but this approach is highly destructive to the dunes habitat and uses a finite resource. Sand fences might also work, but researchers are still analyzing optimum fence designs. Preventing sand dunes from overwhelming cities and agricultural areas has become a priority for the United Nations Environment Program.

One way of preventing sand from accumulating in roadways is planting trees and vegetation along the road.

Examples

- The dunes in the Thar desert in Rajasthan, India.
- Tottori Sand Dunes,Tottori Prefecture, Japan.
- The Kelso Dunes, in the Mojave desert of California.
- Sands of Forvie within the Ythan Estuary complex, Aberdeenshire, Scotland.
- Great Sand Dunes National Park, Colorado, United States.
- Western Sahara.

- White Sands National Monument, U.S.
- Rig-e Jenn in the Central Desert of Iran.
- Rig-e Lut in the Southeast of Iran.
- Indiana Dunes/Sleeping Bear Dunes National Lakeshore, Lake Michigan, U.S.
- Algodones Dunes near Brawley, California, U.S.
- Guadalupe-Nipomo Dunes, Central Coast California, U.S.
- Lencoi Maranhenses in the state of Maranhão, Brazil.
- Mer'eb Dune (also written as Merheb) in United Arab Emirates, used as an arena for motor sports and skiing.
- Monahans Sandhills State Park near Odessa, Texas.
- Oxwich Dunes, near Swansea on the Gower peninsula in Wales.
- Winterton Dunes-Norfolk, England.
- The Killpecker sand dunes of Southwestern Wyoming, U.S.
- Jockey's Ridge State Park—Outer Banks, North Carolina.

Conservation

Dune habitats provide niches for highly specialized plants and animals, including numerous rare and endangered species. Due to human population expansion dunes face destruction through recreation and land development, as well as alteration to prevent encroachment on inhabited areas. Some countries, notably the U.S., New Zealand, Great Britain, Australia, Canada, and the Netherlands have developed extensive programs of dune protection. In the UK, a Biodiversity Action Plan has been developed to assess dunes loss and prevent future dunes destruction.

Extraterrestrial Dunes

Dunes can likely be found in any environment where there is a substantial atmosphere, winds, and dust to be blown. Dunes are common on Mars, and have also been observed in the equatorial regions of Titan by the Cassini probe's radar.

Titan's dunes include large expanses with modal lengths of about 20-30 km. The regions are not topographically confined, resembling sand seas. These dunes are interpreted to be longitudinal dunes whose crests are oriented parallel to the dominant wind direction, which generally indicates west-to-east wind flow. The sand is likely composed of hydrocarbon particles, possibly with some water ice mixed in.

Loess

In some parts of the world, windblown dust and silt blanket the land. This layer of fine, mineral-rich material is called loess.

Loess is mostly created by wind, but can also be formed by glaciers. When glaciers grind rocks to a fine powder, loess can form. Streams carry the powder to the end of the glacier. This sediment becomes loess.

Loess ranges in thickness from a few centimeters to more than 91 meters (300 feet). Unlike other soils, loess is pale and loosely packed. It crumbles easily; in fact, the word "loess" comes from the German word for "loose." Loess is soft enough to carve, but strong enough to stand as sturdy walls. In parts of China, residents build cave-like dwellings in thick loess cliffs.

Extensive loess deposits are found in northern China, the Great Plains of North America, central Europe, and parts of Russia and Kazakhstan. The thickest loess deposits are near the Missouri River in the U.S. state of Iowa and along the Yellow River in China.

Loess accumulates, or builds up, at the edges of deserts. For example, as wind blows across the Gobi, a desert in Asia, it picks up and carries fine particles. These particles include sand crystals made of quartz or mica. It may also contain organic material, such as the dusty remains of skeletons from desert animals.

On the far side of the desert, moisture in the air causes the particles and dust to settle on the ground. There, grass and the roots of other plants trap the dust and hold it to the ground. More dust slowly accumulates, and loess is formed.

Loess often develops into extremely fertile agricultural soil. It is full of minerals and drains water very well. It is easily tilled, or broken up, for planting seeds. Loess usually erodes very slowly—Chinese farmers have been working the loess around the Yellow River for more than a thousand years.

The Loess Hills of western Iowa are more than 61 meters (200 feet) thick.

Properties

Loess is homogeneous, porous, friable, pale yellow or buff, slightly coherent, typically non-stratified and often calcareous. Loess grains are angular with little polishing or rounding and composed of crystals of quartz, feldspar, mica and other minerals. Loess can be described as a rich, dust-like soil.

Loess deposits may become very thick, more than a hundred meters in areas of China and tens of meters in parts of the Midwestern United States. It generally occurs as a blanket deposit that covers areas of hundreds of square kilometers and tens of meters thick.

Loess often stands in either steep or vertical faces. Because the grains are angular, loess will often stand in banks for many years without slumping. This soil has a characteristic called vertical cleavage which makes it easily excavated to form cave dwellings, a popular method of making human habitations in some parts of China. Loess will erode very readily.

In several areas of the world, loess ridges have formed that are aligned with the prevailing winds during the last glacial maximum. These are called "paha ridges" in America and "greda ridges" in Europe. The form of these loess dunes has been explained by a combination of wind and tundra conditions.

Loess near Hunyuan, Shanxi province, China.

Formation

According to Pye, four fundamental requirements are necessary for the formation of loess: a dust source, adequate wind energy to transport the dust, a suitable accumulation area, and a sufficient amount of time.

Periglacial Loess

Periglacial (glacial) loess is derived from the floodplains of glacial braided rivers that carried large volumes of glacial meltwater and sediments from the annual melting of continental icesheets and mountain icecaps during the spring and summer. During the autumn and winter, when melting of the icesheets and icecaps ceased, the flow of meltwater down these rivers either ceased or was greatly reduced. As a consequence, large parts of the formerly submerged and unvegetated floodplains of these braided rivers dried out and were exposed to the wind. Because these floodplains consist of sediment containing a high content of glacially ground flour-like silt and clay, they were highly susceptible to winnowing of their silts and clays by the wind. Once entrained by the wind, particles were then deposited downwind. The loess deposits found along both sides of the Mississippi River Alluvial Valley are a classic example of periglacial loess.

During the Quaternary, loess and loess-like sediments were formed in periglacial environments on mid-continental shield areas in Europe and Siberia, on the margins of high mountain ranges like in Tajikistan and on semi-arid margins of some lowland deserts like in China.

In England, periglacial loess is also known as brickearth.

Non-glacial

Non-glacial loess can originate from deserts, dune fields, playa lakes, and volcanic ash.

Some types of nonglacial loess are:

- Desert loess produced by aeolian attrition of quartz grains.
- Volcanic loess in Ecuador and Argentina.
- Tropical loess in Argentina, Brazil and Uruguay.
- Gypsum loess in Spain.
- Trade wind loess in Venezuela and Brazil.
- Anticyclonic loess in Argentina.

The thick Chinese loess deposits are non-glacial loess having been blown in from deserts in northern China. The loess covering the Great Plains of Nebraska, Kansas, and Colorado is considered to be non-glacial desert loess. Non-glacial desert loess is also found in Australia and Africa.

Fertility

Loess tends to develop into very rich soils. Under appropriate climatic conditions, it is some of the most agriculturally productive terrain in the world.

Soils underlain by loess tend to be excessively drained. The fine grains weather rapidly due to their large surface area, making soils derived from loess rich. One theory states that the fertility of loess soils is due largely to cation exchange capacity (the ability of plants to absorb nutrients from the soil) and porosity (the air-filled space in the soil). The fertility of loess is not due to organic matter content, which tends to be rather low, unlike tropical soils which derive their fertility almost wholly from organic matter.

Even well managed loess farmland can experience dramatic erosion of well over 2.5 kg /m^2 per year. In China the loess deposits which give the Yellow River its color have been farmed and have produced phenomenal yields for over one thousand years. Winds pick up loess particles, contributing to the Asian Dust pollution problem. The largest deposit of loess in the United States, the Loess Hills along the border of Iowa and Nebraska, has survived intensive farming and poor farming practices. For almost 150 years, this loess deposit was farmed with mouldboard ploughs and fall tilled, both intensely erosive. At times it suffered erosion rates of over 10 kilograms per square meter per year. Today this loess deposit is worked as low till or no till in all areas and is aggressively terraced.

Glacial Landforms

Glacial landforms are created by the action of the glacier through the movement of a large ice sheet. The glacial landforms can either be erosional or depositional depending on the action of the glacier. When glaciers retreat leaving behind crashed rocks and debris they create depositional landforms, but if the glaciers expand as a result of their accumulating weight crushing in the process and abrade scoured surface rock or bedrock, then it will lead to the formation of erosional landforms. Depositional landforms include eskers, kame, and Moraine while erosional landforms include Cirque, glacial horns, and arête. Apart from landforms, glaciers may also be striking features including lakes and ponds, particularly in the Polar Regions.

Valley

Valley is an elongate depression of the Earth's surface. Valleys are most commonly drained by rivers and may occur in a relatively flat plain or between ranges of hills or mountains. Those valleys produced by tectonic action are called rift valleys. Very narrow, deep valleys of similar appearance are called gorges. Both of these latter types are commonly cut in flat-lying strata but may occur in other geological situations.

Wherever sufficient rainfall occurs, opportunity exists for the land surface to evolve to the familiar patterns of hills and valleys. There are, of course, hyperarid environments where fluvial activity is minimal. There also are geomorphological settings where the permeability of rocks or sediments induce so much infiltration that water is unable to concentrate on the land surface. Moreover, some landscapes may be so young that insufficient time has elapsed for modification by fluvial action. The role of fluvial action on landscape, including long-term evolutionary processes, is considered here in detail.

Probably the world's deepest sub aerial valley is that of the Kāli Gandaki River in Nepal. Lying between two 8,000-metre (26,000-foot) Himalayan peaks, Dhaulāgiri and Annapūrna, the valley has a total relief of six kilometers (four miles). Because the Himalayas are one of the Earth's most active areas of tectonic uplift, this valley well illustrates the principle that the most rapid down cutting occurs in areas of the most rapid uplift. The reason for this seeming paradox lies in the energetics of the processes of degradation that characterize valley formation. The steeper the gradient or slope of a stream, the greater its expenditure of power on the streambed. Thus, as uplift creates higher relief and steeper slopes, rivers achieve greater power for erosion. As a consequence, the most rapid processes of relief reduction can occur in areas of most rapid relief production.

Perhaps the most famous example of a canyon is the Grand Canyon of the Colorado River in northern Arizona. The Grand Canyon is about 1.6 km (1 mile) deep and 180 meters (590 feet) to 30 km (19 miles) wide and occurs along a 443-km- (275-mile-) long reach where the Colorado River incised into a broad upwarp of sedimentary rocks.

Geomorphic Characteristics

The relief of valleys and canyons is produced by the incising action of rivers. Hillslope processes are indeed critical in the development of valley sides, but it is rivers that lower the level of erosion through degradation. Rivers ultimately adjust to a base level, defined as the lowest point at which potential energy can be transformed to the kinetic energy of river flow. In most cases, the ultimate base level for rivers is sea level. Some rivers drain to enclosed basins below sea level, as, for example, the Jordan River, which flows to the Dead Sea in Israel and Jordan. Moreover, rivers may adjust to local base levels, including zones of resistance to incision, lakes, and dams (both natural and artificial).

Valley Longitudinal Profiles

The longitudinal profile of a valley is the gradient throughout its length. Valleys formed by river action typically have a concave upward profile, steep in the headwaters and gentle in the lower reaches. The lower end of such a profile is adjusted to an effective lower limit of erosion defined by the base level.

In an ideal case of river adjustment to uniformly resistant materials, the longitudinal profile of a stream assumes a characteristic form that minimizes variations in transporting power. Power in a river derives from the rate of transfer of potential energy, dE/dt, which depends on the rate of fall in elevation of water, dy/dt, according to,

$$\frac{dE}{dt} = mg\frac{dy}{dt},$$

where E is energy, t is time, m is mass, g is the acceleration of gravity, and y is elevation. The rate of fall in elevation, in turn, can be expressed as follows:

$$\frac{dy}{dt} = \frac{dy}{dx}\frac{dx}{dt} = SV$$

where S is the slope (fall in elevation, dy, with downstream horizontal distance, dx) and V is the flow velocity (change in horizontal distance, dx, with time, dt).

Combining equations (1) and (2) and using the fluid density ρ (mass per unit volume of water), one obtains,

$$\frac{dE}{dt} = pg(W \cdot D \cdot L)SV,$$

where W is channel width, D is channel depth, L is a unit length of stream, and the other parameters are as defined above. Because flow discharge Q is defined as:

$$Q = W \cdot D \cdot V,$$

the power per unit length of flow, Ω, can be expressed as:

$$\Omega = dE / dt / L = pgQS.$$

It should be noted that in order to minimize variation in power, a river increasing its discharge in a downstream direction must decrease its slope. Thus, slope must be constantly decreasing downstream, explaining the concave upward character of the longitudinal profile.

The idealized concave upward longitudinal profile defined purely by energy considerations, only occurs where channel bed resistances and adequate adjustment time permit. Resistant zones of bedrock require greater power for a stream to incise at a given discharge Q than do less resistant zones. Therefore, by equation $\Omega = dE / dt / L = pgQS$. the stream gradient S must be locally steeper at resistant zones. Similarly, a rapid base-level change, such as a fall of sea level, may not allow adequate time for the entire longitudinal profile to adjust. One indication of such effects on a longitudinal profile is a nick point, or abrupt change in slope of the profile.

Valley Cross Profiles

The cross profiles of valleys involve a combination of fluvial and hillslope processes. Although slopes and rivers are often studied separately by process geomorphologists, hills and valleys are the features that dominate landscapes. In upland areas cross profiles of valleys are often narrow and deep. Canyon morphologies are most common. Further downstream, valley floors are wider and often dominated by floodplains and terraces.

Types of Valleys

One of the few classifications of valleys is that used by the German climatic geomorphologists Herbert Louis and Julius Büdel. In areas of rapid uplift and intense fluvial action such as tropical mountains, Kerbtal forms occur. These are characterized by steep, knife-edge ridges and valley slopes meeting in a V-shape. Where slopes are steep but a broad valley floor occurs, Sohlenkerbtal (meaning precisely a valley with such characteristics) is the prevailing form. Valleys of this kind develop under the influence of groundwater flow in Hawaii. Gutter-shaped valleys with convex sides and broad floors are called Kehltal; and broad, flat valleys of planation surfaces are termed Fachmuldental.

It is important to remember that the form of valleys reflects not only modern processes but also ancient ones. The entire valley or some landforms within it may be relict, with features inherited from past geologic periods during which occurred tectonic and climatic processes of intensities quite different from those prevailing today.

Hillslopes

Hillslopes constitute the flanks of valleys and the margins of eroding uplands. They are the major zones where rock and soil are loosened by weathering processes and then transported down gradient, often to a river channel.

Two major varieties of hillslopes occur in nature. On weathering-limited slopes, transport processes are so efficient that debris is removed more quickly than it can be generated by further weathering. Such hillslopes develop a faceted or angular morphology in which an upper free face, or cliff, contributes debris to a lower slope of accumulation. Slopes of this sort are especially common on bare rock where the profile of the slope is determined by the resistance of the rock, not by the erosional processes acting on it. One consequence of this is that many rock slopes retreat parallel to themselves in order to preserve the characteristic slope angle for a rock type of given strength. If the features of the rock change with depth into the slope, however, the characteristic angle of the slope will change. Rock slopes develop where weathering and soil erosion are slow (as in arid regions) and where rock resistance is high.

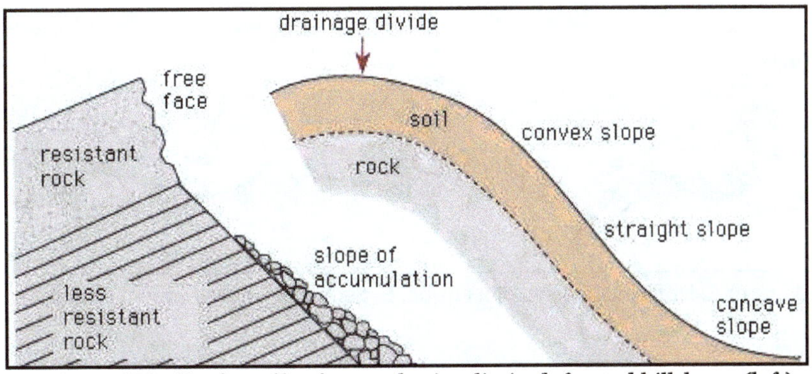

Comparison of idealized profiles for weathering-limited, faceted hillslopes (left) and transport-limited, sigmoid hillslopes (right).

The second major variety of slope is transport limited. Transport-limited slopes occur where weathering processes are efficient at producing debris but where transport processes are inefficient at removing it from the slope. Such slopes lack free faces and faceted appearances, and they are generally covered with a soil mantle. The profile of this type of slope generally has a sigmoid appearance, with convex, straight, and concave segments. The shape of the slope is an expression of the process acting upon it.

Convex slope segments commonly occur in the upper parts of soil-mantled slopes, as near the drainage divide. G.K. Gilbert elucidated the principles applying to convex slopes in his study of piles of mining-waste debris in California. The processes of soil creep and raindrop splash erode soil on the upper parts of slopes. Since soil eroded from the upper slope must pass each point below it, the volume of soil moved increases with distance from the divide. Since the transport rate for creep and rain splash is proportional to the slope angle, the slope angle must also increase from the divide, resulting in the slope convexity.

Straight slope segments are dominated by mass movement processes. Talus slopes are a type in which debris piles up to a characteristic angle of repose. When new debris is added to the slope, thereby locally increasing the angle, the slope adjusts by movement of the debris to reestablish the angle. Again, the result is a dynamic equilibrium in which the landform adjusts to processes acting upon it.

Concave slopes are especially common where overland-flow runoff transports sediment derived from upper slopes. Because the collection area for wash increases downslope and discharge Q is proportional to collection area, stream power—equation $\Omega = dE / dt / L = pgQS$.—can be maintained at lower slope angles. In addition, the size of particles being transported decreases downslope because of weathering and abrasion. Because the finer particles are easier to transport, slope angles can be reduced in the downslope direction. The result is a concave shape to the slope profile.

Erosional Landforms

Cirque/Corrie

Is an amphitheatre-shaped hollow basin cut into a mountain ridge. It has steep sided slope on three sides, an open end on one side and a flat bottom. When the ice melts, the cirque may develop into a tarn lake (figure).

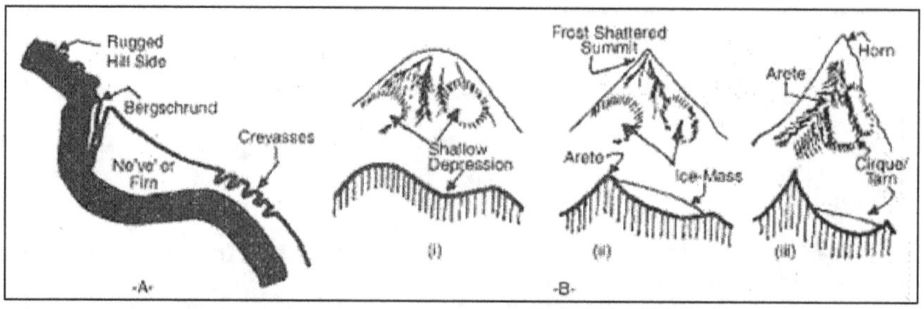

Figure: A. Different aspects of a cirque. B. Sequential development of cirque.

Glacial Trough

Glacial Trough is an original stream-cut valley, further modified by glacial action. Step- formation takes place at maturity; otherwise it is an ungraded and irregular feature.

'U' Shaped Valley

'U' Shaped Valley is another typically glacial feature. Since glacial mass is heavy and slow moving, erosional activity is uniform—horizontally as well as vertically. A steep sided and flat-bottomed valley results, which has a 'U' shaped profile.

Hanging Valley

Hanging Valley is formed when smaller tributaries are unable to cut as deeply as bigger ones and remain 'hanging' at higher levels than the main valley as discordant tributaries. This may happen due to glacial tilting or faulting.

Arete

Arete is a steep-sided, sharp-tipped summit with the glacial activity cutting into it from two sides.

Horn is a ridge that acquires a 'horn' shape when the piedmont glacier surrounds a summit.

Fjord is formed as a steep-sided narrow entrance-like feature at the coast where the stream meets the coast. Fjords are common in Norway, Greenland and New Zealand.

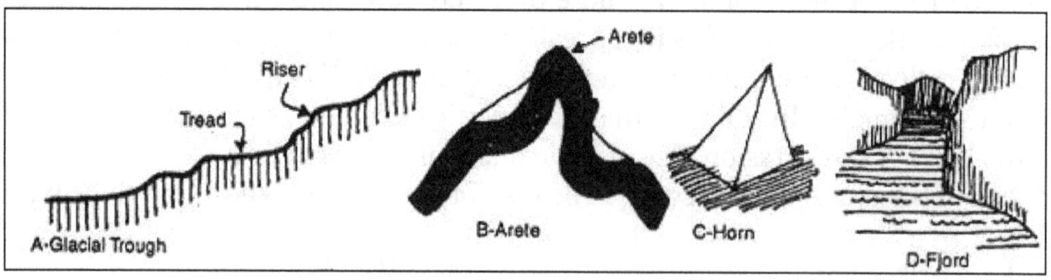

Figure: Landforme formed by glacial arosion.

Mammillated Field

Mammillated Field is a term used for an ice-eroded field.

Depositional Landforms

Outwash Plain:

When the glacier reaches its lowest point and melts, it leaves behind a stratified deposition material, consisting of rock debris, clay, sand, gravel etc. This layered surface is called till plain or an outwash plain and a downward extension of the deposited clay material is called valley train (figure).

Esker:

Is a winding ridge of unassorted depositions of rock, gravel, clay etc. running along a glacier in a till plain. The eskers resemble the features of an embankment and are often used for making roads. If the melting of glacier has been punctuated, it is reflected in a local widening of the esker and here it is called a beaded esker (figure).

Kame:

Terraces are the broken ridges or unassorted depositions looking like hummocks in a till plain (figure).

Drumlin:

Is an inverted boat-shaped deposition in a till plain caused by deposition. The erosional counterpart is called a roche moutonne (figure).

Kettle Holes:

Can be formed when the deposited material in a till plain gets depressed locally and forms a basin (figure).

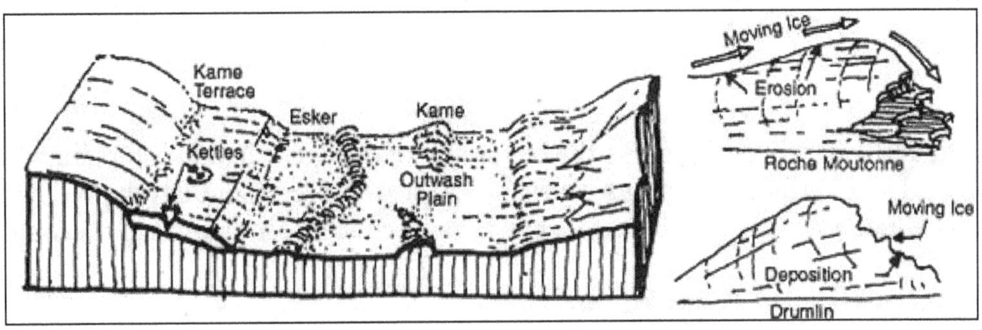

Figure: Landforme formed by glacial deposition.

Moraine:

Moraine is a general term applied to rock fragments, gravel, sand, etc., carried by a glacier. Depending on its position, the moraine can be ground, lateral, medial or terminal moraine. The material dropped at the end of a valley glacier in the form of a ridge is called the terminal moraine. Each time a glacier retreats, a fresh terminal moraine is left at a short distance behind the first one.

The material deposited at either of its sides is known as lateral moraine. When two glaciers join, their lateral moraines also join near their confluence and are called medial moraines. Many Alpine pastures in the Himalayas like the Margs of Kashmir occupy the sites of morainic deposits of old river valleys. The excessive load that cannot be carried forward by a glacier is deposited on its own bed or at the base and appears as what is known as ground moraine. (figure below)

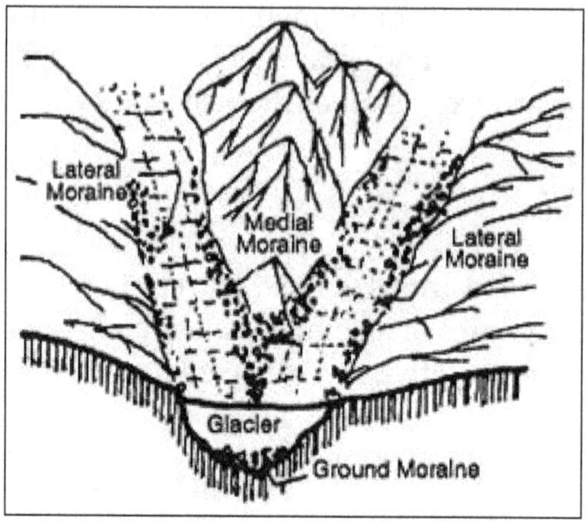

Figure: Different positions of moraines.

Glacial Cycle of Erosion

Youth:

The stage is marked by the inward cutting activity of ice in a cirque. Aretes and horns are emerging. The hanging valleys are not prominent at this stage.

Maturity:

The valley glacier gets transformed into trunk glacier and hanging valleys start emerging. The opposite cirques come closer and the glacial trough acquires a stepped profile which is regular and graded.

Old Age:

Emergence of a 'U'-shaped valley marks the beginning of old age. An outwash plain with features such as eskers, kame terraces, drumlins, kettle holes etc. is a prominent development. The opposite cirques coalesce and the summit heights are greatly reduced. Mountain tops become rounded.

Other Types of Glacial Landforms

Mesa

A mesa is a flat-topped mountain or hill. It is a wide, flat, elevated landform with steep sides.

Mesa is a Spanish word that means table. Spanish explorers of the American southwest, where many mesas are found, used the word because the tops of mesas look like the tops of tables.

Mesas are formed by erosion, when water washes smaller and softer types of rocks away from the top of a hill. The strong, durable rock that remains on top of a mesa is called cap rock. A mesa is usually wider than it is tall.

Mesas are usually found in dry regions where rock layers are horizontal. The Grand Mesa in the U.S. state of Colorado, considered the largest mesa in the world, has an area of about 1,300 square kilometers (500 square miles) and stretches for 64 kilometers (40 miles).

This gorgeous mesa towers above the surrounding vegetation of the Guiana Highlands of South America.

Formation

Har Qatum, a mesa located on the southern edge of Makhtesh Ramon, Israel.

Mesas form by weathering and erosion of horizontally layered rocks that have been uplifted by tectonic activity. Variations in the ability of different types of rock to resist weathering and erosion cause the weaker types of rocks to be eroded away, leaving the more resistant types of rocks topographically higher than their surroundings. This process is called differential erosion. The most resistant rock types include sandstone, conglomerate, quartzite, basalt, chert, limestone, lava flows and sills. Lava flows and sills, in particular, are very resistant to weathering and erosion, and often form the flattop, or caprock, of a mesa. The less resistant rock layers are mainly made up of shale, a softer rock that weathers and erodes more easily.

The differences in strength of various rock layers is what gives mesas their distinctive shape. Less resistant rocks are eroded away on the surface into valleys, where they collect water drainage from the surrounding area, while the more resistant layers are left standing out. A large area of very resistant rock, such as a sill may shield the layers below it from erosion while the softer rock surrounding it is eroded into valleys, thus forming a caprock.

Differences in rock type also reflect on the sides of a mesa, as instead of smooth slopes, the sides are broken into a staircase pattern called "cliff-and-bench topography". The more resistant layers form the cliffs, or stairsteps, while the less resistant layers form gentle slopes, or benches, between the cliffs. Cliffs retreat and are eventually cut off from the main cliff, or plateau, by basal sapping. When the cliff edge does not retreat uniformly, but instead is indented by headward eroding streams, a section can be cut off from the main cliff, forming a mesa.

Basal sapping occurs as water flowing around the rock layers of the mesa erodes the underlying soft shale layers, either as surface runoff from the mesa top or from groundwater moving through permeable overlying layers, which leads to slumping and flowage of the shale. As the underlying shale erodes away, it can no longer support the overlying cliff layers, which collapse and retreat. When the caprock has caved away to the point where only a little remains, it is known as a butte.

Canyon

A canyon is a deep, narrow valley with steep sides. "Canyon" comes from the Spanish word canon, which means "tube" or "pipe." The term "gorge" is often used to mean "canyon," but a gorge is almost always steeper and narrower than a canyon.

The movement of rivers, the processes of weathering and erosion, and tectonic activity create canyons.

River Canyons

The most familiar type of canyon is probably the river canyon. The water pressure of a river can cut deep into a river bed. Sediments from the river bed are carried downstream, creating a deep, narrow channel.

Rivers that lie at the bottom of deep canyons are known as entrenched rivers. They are entrenched because, unlike rivers in wide, flat flood plains, they do not meander and change their course.

The Yarlung Zangbo Grand Canyon in Tibet, a region of southwestern China, was formed over millions of years by the Yarlung Zangbo River. This canyon is the deepest in the world—at some points extending more than 5,300 meters (17,490 feet) from top to bottom. Yarlung Zangbo Canyon is also one of the world's longest canyons, at about 500 kilometers (310 miles).

Weathering and Erosion

Weathering and erosion also contribute to the formation of canyons. In winter, water seeps into cracks in the rock. This water freezes. As water freezes, it expands and turns into ice. Ice forces the cracks to become larger and larger, eroding bits of stone in the process. During brief, heavy rains, water rushes down the cracks, eroding even more rocks and stone. As more rocks crumble and fall, the canyon grows wider at the top than at the bottom.

When this process happens in soft rock, such as sandstone, it can lead to the development of slot canyons. Slot canyons are very narrow and deep. Sometimes, a slot canyon can be less than a meter (3 feet) wide, but hundreds of meters deep. Slot canyons can be dangerous. Their sides are usually very smooth and difficult to climb.

Some canyons with hard, underlying rock may develop cliffs and ledges after their softer, surface rock erodes. These ledges look like giant steps.

Sometimes, entire civilizations can develop on and around these canyon ledges. Native American nations, such as the Hopi and Sinagua, made cliff dwellings. Cliff dwellings were apartment-style shelters that housed hundreds of people. The shaded, elevated ledges in Walnut Canyon and Canyon de Chelly, in Arizona, provided protection from hostile neighbors and the burning desert sun.

Hard-rock canyons that are open at one end are called box canyons. The Hopi and Navajo people often used box canyons as natural corrals for sheep and cattle. They simply built a gate on the open side of the box canyon, and closed it when the animals were inside.

Limestone is a type of hard rock often found in canyons. Sometimes, limestone erodes and forms caves beneath the earth. As the ceilings of these caves collapse, canyons form. The Yorkshire Dales, an area in northern England, is a collection of river valleys and canyons created by limestone cave collapses.

Tectonic Uplift

Canyons are also formed by tectonic activity. As tectonic plates beneath the Earth's crust shift and collide, their movement can change the area's landscape. Sometimes, tectonic activity causes an area of the Earth's crust to rise higher than the surrounding land. This process is called tectonic uplift. Tectonic uplift can create plateaus and mountains. Rivers and glaciers that cut through these elevated areas of land create deep canyons.

The Grand Canyon, in the U.S. state of Arizona, is a product of tectonic uplift. The Grand Canyon, up to 447 kilometers (277 miles) long, 29 kilometers (18 miles) wide, and 1.8 kilometers (6,000 feet) deep, is the largest canyon in the United States. The Grand Canyon has been carved, over millions of years, as the Colorado River cuts through the Colorado Plateau. The Colorado Plateau is a large area that was elevated through tectonic uplift millions of years ago. Geologists debate the age of the canyon itself–it may be between 5 million and 70 million years old.

Canyons are like silent journals of an area's history over thousands or even millions of years. By studying the exposed layers of rock in a canyon wall, experts can learn about how the climate changed, what kind of organisms were alive at certain times, and perhaps even how the canyon may change in the future.

For example, geologists studying layers of rock in the Columbia River Gorge, in the U.S. states of Washington and Oregon, discovered that the oldest rocks there are at least 17 million years old. They also found out the rocks are dark-black basalt, made from hardened lava. From this, geologists determined that the rocks formed when volcanoes erupted and their lava spilled out onto the land. Over millions of years, the Columbia River and Ice Age glaciers cut through the area and exposed its volcanic beginnings.

Canyons are also important to paleontology, or the study of fossils. Fossils are often best preserved in dry, hot areas. Since canyons usually form under the same conditions, they are good places to examine fossils.

The layers of sediment revealed by a canyon can make it easier to date fossils. For example, a new area of dinosaur tracks was discovered in the U.S. state of Utah at Glen Canyon National

Recreation Area in 2009. These tracks reveal new information about a group of dinosaurs called ornithopods. Paleontologists analyzed the layers of rock surrounding the fossils to estimate how old they were. These new dinosaur tracks show that ornithopods were alive 20 million years earlier than scientists thought.

Geologists study canyons to determine how the landscape will change in the future. The erosion patterns and thickness of different layers can reveal the climate during different years. A series of very dry years will have very thin layers of rock, when little erosion took place. The overall pattern of erosion and layering reveals the rate of water flow, from both the river and rain, through a canyon.

Geologists estimate that the Grand Canyon, for example, is being eroded at a rate of 0.3 meters (1 foot) every 200 years. The Colorado Plateau, the geologic area where the Grand Canyon is located, is a very stable area. Geologists expect the Grand Canyon to continue to deepen as long as the Colorado River flows.

Submarine Canyons

Some of the deepest canyons lie beneath the ocean. These submarine canyons cut into continental shelves and continental slopes—the edges of continents that are underwater.

Some submarine canyons were carved by rivers that flowed during periods when the sea level was lower, and the continental shelves were exposed. The Hudson Canyon extends 750 kilometers (450 miles) into the Atlantic Ocean, from the mouth of the Hudson River, in the U.S. states of New York and New Jersey. At least part of the Hudson Canyon was the river bed during the last ice age, when sea levels were much lower.

Submarine canyons can also develop when powerful ocean currents sweep away sediments. Just as rivers erode land, these currents carve deep canyons in the ocean floor. Strong currents of the Atlantic Ocean prevent Whittard Canyon, about 400 kilometers (248 miles) south of the coast of Ireland, from filling with sediment. Scientists studying Whittard Canyon believe glacial water mixed with seawater to rush into the submarine canyon thousands of years ago.

The formation of some submarine canyons is still a mystery. Monterey Canyon is a deep submarine canyon off the coast of the U.S. state of California. It has been compared to the Grand Canyon because of its size. It is 152 kilometers (95 miles) long and 3.2 kilometers (2 miles) deep at its deepest point. Geologists still aren't certain how Monterey Canyon was formed. One theory is that the canyon was formed by an ancient outlet of the Sacramento or Colorado Rivers. Another theory is that it was formed by tectonic activity—an earthquake splitting apart the rock with enormous force. Scientists believe the canyon was formed 25 million to 30 million years ago.

The depth of submarine canyons makes them hard to explore. Scientists usually use remotely operated vehicles (ROVs) to conduct studies. Sometimes, they can use a submersible, a special kind of submarine. The Monterey Bay Aquarium Research Institute (MBARI) uses a vehicle called Ventana to explore Monterey Canyon. Through the Ventana and other research vehicles, MBARI scientists have discovered new species of organisms living in the canyon, from tiny sea anemones to giant squid.

The Grand Canyon, brought to you by the Colorado Plateau and the Colorado River.

Fluvial and Coastal Landforms

Coastal landforms are any of the relief features present along any coast, the result of a combination of processes, sediments, and the geology of the coast itself.

The coastal environment of the world is made up of a wide variety of landforms manifested in a spectrum of sizes and shapes ranging from gently sloping beaches to high cliffs, yet coastal landforms are best considered in two broad categories: erosional and depositional. In fact, the overall nature of any coast may be described in terms of one or the other of these categories. However, that each of the two major landform types may occur on any given reach of coast.

Factors and Forces in the Formation of Coastal Features

The landforms that develop and persist along the coast are the result of a combination of processes acting upon the sediments and rocks present in the coastal zone. The most prominent of these processes involves waves and the currents that they generate, along with tides. Other factors that significantly affect coastal morphology are climate and gravity.

Waves

The most obvious of all coastal processes is the continual motion of the waves moving toward the beach. Waves vary considerably in size over time at any given location and also vary markedly from place to place. Waves interact with the ocean bottom as they travel into shallow water; as a result, they cause sediment to become temporarily suspended and available for movement by coastal currents. The larger the wave, the deeper the water in which this process takes place and the larger the particle that can be moved. Even small waves that are only a few tens of centimeters high can pick up sand as they reach the shore. Larger waves can move cobbles and rock material as large as boulders.

Generally, small waves cause sediment—usually sand—to be transported toward the coast and to become deposited on the beach. Larger waves, typically during storms, are responsible for the removal of sediment from the coast and its conveyance out into relatively deep water.

Waves erode the bedrock along the coast largely by abrasion. The suspended sediment particles in waves, especially pebbles and larger rock debris, have much the same effect on a surface as sandpaper does. Waves have considerable force and so may break up bedrock simply by impact.

Longshore Currents

Waves usually approach the coast at some acute angle rather than exactly parallel to it. Because of this, the waves are bent (or refracted) as they enter shallow water, which in turn generates a current along the shore and parallel to it. Such a current is called a longshore current, and it extends from the shoreline out through the zone of breaking waves. The speed of the current is related to the size of the waves and to their angle of approach. Under rather quiescent conditions, longshore currents move only about 10–30 centimeters per second; however, under stormy conditions they may exceed one meter per second. The combination of waves and longshore current acts to transport large quantities of sediment along the shallow zone adjacent to the shoreline.

Because longshore currents are caused by the approaching and refracting waves, they may move in either direction along the coast, depending on the direction of wave approach. This direction of approach is a result of the wind direction, which is therefore the ultimate factor in determining the direction of longshore currents and the transport of sediment along the shoreline.

Although a longshore current can entrain sediment if it moves fast enough, waves typically cause sediment to be picked up from the bottom, and the longshore current transports it along the coast. In some locations there is quite a large volume of net sediment transport along the coast because of a dominance of one wind direction—and therefore wave direction—over another. This volume may be on the order of 100,000 cubic meters per year. Other locations may experience more of a balance in wave approach, which causes the longshore current and sediment transport in one direction to be nearly balanced by the same process in the other direction.

Rip Currents

Another type of coastal current caused by wave activity is the rip current (incorrectly called rip tide in popular usage). As waves move toward the beach, there is some net shoreward transport of water. This leads to a slight but important upward slope of the water level (setup), so that the absolute water level at the shoreline is a few centimeters higher than it is beyond the surf zone. This situation is an unstable one, and water moves seaward through the surf zone in an effort to relieve the instability of the sloping water. The seaward movement is typically confined to narrow pathways. In most cases, rip currents are regularly spaced and flow at speeds of up to several tens of centimeters per second. They can carry sediment and often are recognized by the plume of suspended sediment moving out through the surf zone. In some localities rip currents persist for months at the same site, whereas in others they are quite ephemeral.

Tides

The rise and fall of sea level caused by astronomical conditions is regular and predictable. There is a great range in the magnitude of this daily or semi-daily change in water level. Along some coasts the tidal range is less than 0.5 meter, whereas in the Bay of Fundy in southeastern Canada the maximum tidal range is just over 16 meters. A simple but useful classification of coasts is based solely on tidal range without regard to any other variable. Three categories have been established: micro-tidal (less than two meters), meso-tidal (two to four meters), and macro-tidal (more than four meters). Micro-tidal coasts constitute the largest percentage of the world's coasts, but the other two categories also are widespread.

The role of tides in molding coastal landforms is twofold: (1) tidal currents transport large quantities of sediment and may erode bedrock, and (2) the rise and fall of the tide distributes wave energy across a shore zone by changing the depth of water and the position of the shoreline.

Tidal currents transport sediment in the same way that longshore currents do. The speeds necessary to transport the sediment (typically sand) are generated only under certain conditions—usually in inlets, at the mouths of estuaries, or any other place where there is a constriction in the coast through which tidal exchange must take place. Tidal currents on the open coast, such as along a beach or rocky coast, are not swift enough to transport sediment. The speed of tidal currents in constricted areas, however, may exceed two meters per second, especially in inlets located on a barrier island complex. The speed of these tidal currents is dictated by the volume of water that must pass through the inlet during a given flood or ebb-tide cycle. This may be either six or 12 hours in duration, depending on whether the local situation is semidiurnal (12-hour cycle) or diurnal (24-hour cycle). The volume of water involved, called the tidal prism, is the product of the tidal range and the area of the coastal bay being served by the inlet. This means that though there may be a direct relationship between tidal range and tidal-current speed, it is also possible to have very swift tidal currents on a coast where the tidal range is low if the bay being served by the inlet is quite large. This is a very common situation along the coast of the Gulf of Mexico where the range is typically less than one meter but where there are many large coastal bays.

The rise and fall of the tide along the open coast has an indirect effect on sediment transport, even though currents capable of moving sediment are not present. As the tide comes in and then retreats along a beach or on a rocky coast, it causes the shoreline to move accordingly. This movement of the shoreline changes the zone where waves and longshore currents can do their work. Tidal range in combination with the topography of the coast is quite important in this situation. The greater the tidal range, the more effect this phenomenon has on the coast. The slope of a beach or other coastal landform also is important, however, because a steep cliff provides only a nominal change in the area over which waves and currents can do their work even in a macro-tidal environment. On the other hand, a broad, gently sloping beach or tidal flat may experience a change in the shoreline of as much as one kilometer during a tidal cycle in a macro-tidal setting. Examples of this situation occur in the Bay of Fundy and along the West German coast of the North Sea.

Other Factors and Processes

Climate is an extremely important factor in the development of coastal landforms. The elements of climate include rainfall, temperature, and wind.

Rainfall is important because it provides runoff in the form of streams and also is a factor in producing and transporting sediment to the coast. This fact gives rise to a marked contrast between the volume and type of sediment carried to the coast in a tropical environment and those in a desert environment.

Temperature is important for two quite different reasons. It is a factor in the physical weathering of sediments and rocks along the coast and in the adjacent drainage basins. This is particularly significant in cold regions where the freezing of water within cracks in rocks causes the rocks to fragment and thereby yield sediment. Some temperate and arctic regions have shore ice up to several months each year. Under these conditions there is no wave impact, and the coast becomes essentially static until the ice thaws or breaks up during severe storms. Such conditions prevail for three to four months along much of the coast of the Great Lakes in North America.

Wind is important primarily because of its relationship to waves. Coasts that experience prolonged and intense winds also experience high wave-energy conditions. Seasonal patterns in both wind direction and intensity can be translated directly into wave conditions. Wind also can be a key factor in directly forming coastal landforms, particularly coastal dunes. The persistence of onshore winds throughout much of the world's coast gives rise to sand dunes in all places where enough sediment is available and where there is a place for it to accumulate.

Gravity, too, plays a major role in coastal processes. Not only is it indirectly involved in processes associated with wind and waves but it also is directly involved through downslope movement of sediment and rock as well. This role is particularly evident along shoreline cliffs where waves attack the base of the cliffs and undercut the slope, resulting in the eventual collapse of rocks into the sea or their accumulation as debris at the base of the cliffs.

Landforms Of Erosional Coasts

There are two major types of coastal morphology: one is dominated by erosion and the other by deposition. They exhibit distinctly different landforms, though each type may contain some features of the other. In general, erosional coasts are those with little or no sediment, whereas depositional coasts are characterized by abundant sediment accumulation over the long term. Both temporal and geographic variations may occur in each of these coastal types.

Erosional coasts typically exhibit high relief and rugged topography. They tend to occur on the leading edge of lithospheric plates, the west coasts of both North and South America being excellent examples. Glacial activity also may give rise to erosional coasts, as in northern New England and in the Scandinavian countries. Typically, these coasts are dominated by exposed bedrock with steep slopes and high elevations adjacent to the shore. Although these coasts are erosional, the rate of shoreline retreat is slow due to the resistance of bedrock to erosion. The type of rock and its lithification are important factors in the rate of erosion.

Sea Cliffs

The most widespread landforms of erosional coasts are sea cliffs. These very steep to vertical bedrock cliffs range from only a few meters high to hundreds of meters above sea level. Their vertical nature is the result of wave-induced erosion near sea level and the subsequent collapse of rocks at higher elevation. Cliffs that extend to the shoreline commonly have a notch cut into them where waves have battered the bedrock surface.

At many coastal locations there is a thin, narrow veneer of sediment forming a beach along the base of sea cliffs. This sediment may consist of sand, but it is more commonly composed of coarse material—cobbles or boulders. Beaches of this kind usually accumulate during relatively low wave-energy conditions and are removed during the stormy season when waves are larger. The coasts of California and Oregon contain many places where this situation prevails. The presence of even a narrow beach along a rocky coast provides the cliffs protection against direct wave attack and slows the rate of erosion.

Wave-cut Platforms

At the base of most cliffs along a rocky coast one finds a flat surface at about the mid-tide elevation. This is a bench like feature called a wave-cut platform, or wave-cut bench. Such surfaces may measure from a few meters to hundreds of meters wide and extend to the base of the adjacent cliff. They are formed by wave action on the bedrock along the coast. The formation process can take a long time, depending on the type of rock present. The existence of extensive wave-cut platforms thus implies that sea level did not fluctuate during the periods of formation. Multiple platforms of this type along a given reach of coast indicate various positions of sea level.

Sea Stacks

Erosion along rocky coasts occurs at various rates and is dependent both on the rock type and on the wave energy at a particular site. As a result of the above-mentioned conditions, wave-cut platforms may be incomplete, with erosional remnants on the horizontal wave-cut surface. These remnants are called sea stacks, and they provide a spectacular type of coastal landform. Some are many meters high and form isolated pinnacles on the otherwise smooth wave-cut surface. Because erosion is a continual process, these features are not permanent and will eventually be eroded, leaving no trace of their existence.

Sea Arches

Another type of erosional landform is the sea arch, which forms as the result of different rates of erosion typically due to the varied resistance of bedrock. These archways may have an arcuate or rectangular shape, with the opening extending below water level. The height of an arch can be up to tens of meters above sea level.

It is common for sea arches to form when a rocky coast undergoes erosion and a wave-cut platform develops. Continued erosion can result in the collapse of an arch, leaving an isolated sea stack on the platform. Still further erosion removes the stack, and eventually only the wave-cut platform remains adjacent to the eroding coastal cliff.

Landforms Of Depositional Coasts

Coasts adjacent to the trailing edge of lithospheric plates tend to have widespread coastal plains and low relief. The Atlantic and Gulf coasts of the United States are representative. Such coasts may have numerous estuaries and lagoons with barrier islands or may develop river deltas. They are characterized by an accumulation of a wide range of sediment types and by many varied coastal environments. The sediment is dominated by mud and sand; however, some gravel may be present, especially in the form of shell material.

Depositional coasts may experience erosion at certain times and places due to such factors as storms, depletion of sediment supply, and rising sea level. The latter is a continuing problem as the mean annual temperature of the Earth rises and the ice caps melt. Nevertheless, the overall, long-range tendency along these coasts is that of sediment deposition.

Waves, wave-generated currents, and tides significantly influence the development of depositional landforms. In general, waves exert energy that is distributed along the coast essentially parallel to it. This is accomplished by the waves themselves as they strike the shore and also by the longshore currents that move along it. In contrast, tides tend to exert their influence perpendicular to the coast as they flood and ebb. The result is that the landforms that develop along some coasts are due primarily to wave processes while along other coasts they may be due mainly to tidal processes. Some coasts are the result of near equal balance between tide and wave processes. As a consequence, investigators speak of wave-dominated coasts, tide-dominated coasts, and mixed coasts.

A wave-dominated coast is one that is characterized by well-developed sand beaches typically formed on long barrier islands with a few widely spaced tidal inlets. The barrier islands tend to be narrow and rather low in elevation. Longshore transport is extensive, and the inlets are often small and unstable. Jetties are commonly placed along the inlet mouths to stabilize them and keep them open for navigation. The Texas and North Carolina coasts of the United States are excellent examples of this coastal type.

Tide-dominated coasts are not as widespread as those dominated by waves. They tend to develop where tidal range is high or where wave energy is low. The result is a coastal morphology that is dominated by funnel-shaped embayments and long sediment bodies oriented essentially perpendicular to the overall coastal trend. Tidal flats, salt marshes, and tidal creeks are extensive. The West German coast of the North Sea is a good example of such a coast.

Mixed coasts are those where both tidal and wave processes exert considerable influence. These coasts characteristically have short stubby barrier islands and numerous tidal inlets. The barriers commonly are wide at one end and narrow at the other. Inlets are fairly stable and have large sediment bodies on both their landward and seaward sides. The Georgia and South Carolina coasts of the United States typify a mixed coast.

General Coastal Morphology

Depositional coasts can be described in terms of three primary large-scale types: (1) deltas, (2) barrier island/estuarine systems, and (3) strand-plain coasts. The latter two have numerous features in common.

Barrier Island/Estuarine Systems

Many depositional coasts display a complex of environments and landforms that typically occur together. Irregular coasts have numerous embayments, many of which are fed by streams. Such embayments are called estuaries, and they receive much sediment due to runoff from an adjacent coastal plain. Seaward of the estuaries are elongate barrier islands that generally parallel the shore. Consisting mostly of sand, they are formed primarily by waves and longshore currents. These barrier islands are typically separated from the mainland and may have lagoons, which are long, narrow, coastal bodies of water situated between the barrier and the mainland.

Most barrier islands contain a well-developed beach, coastal dunes, and various environments on their landward side, including tidal flats, marshes, or wash over fans. Such coastal barriers are typically interrupted by tidal inlets, which provide circulation between the various coastal bays and the open marine environment. These inlets also are important pathways for organisms that migrate between coastal and open marine areas as well as for pleasure and commercial boat traffic.

Strand-plain Coasts

Some wave-dominated coasts do not contain estuaries and have no barrier island system. These coasts, however, do have beaches and dunes, and may even have coastal marshes. The term strand plain has been applied to coasts of this sort. Examples include parts of western Louisiana and eastern Texas. In most respects, they are similar in morphology to barrier islands but lack inlets.

Beaches and Coastal Dunes

There are several specific landforms representative of coastal environments that are common to each of the three major categories described above. Especially prominent among these are beaches and dunes. They are the primary landforms on barrier islands, strand-plain coasts, and many deltas, particularly the wave-dominated variety.

Beaches

A consideration of the beach must also include the seaward adjacent near shore environment because the two are intimately related. The near shore environment extends from the outer limit of the longshore bars that are usually present to the low-tide line. In areas where longshore bars are absent, it can be regarded as coincident with the surf zone. The beach extends from the low-tide line to the distinct change in slope and/or material landward of the unvegetated and active zone of sediment accumulation. It may consist of sand, gravel, or even mud, though sand is the most common beach material.

The beach profile typically can be divided into two distinct parts: (1) the seaward and relatively steep sloping foreshore, which is essentially the intertidal beach, and (2) the landward, nearly horizontal backshore. Beach profiles take on two different appearances, depending on conditions at any given time. During quiescent wave conditions, the beach is said to be accretional, and both the foreshore and backshore are present. During storm conditions, however, the beach experiences erosion, and the result is typically a profile that shows only the seaward sloping foreshore. Because the beach tends to repair itself during nonstorm periods, a cyclic pattern of profile shapes is common.

The near shore zone is where waves steepen and break, and then re-form in their passage to the beach, where they break for the last time and surge up the foreshore. Much sediment is transported in this zone, both along the shore and perpendicular to it. During storms the waves tend to be steep, and erosion of the beach occurs with sediment transported offshore. The intervening calmer conditions permit sediment to be transported landward and rebuild the beach. Because wave conditions may change daily, the nature of the profile and the sediment on the foreshore portion of the beach may also change daily. This is the zone of continual change on the beach.

The backshore of the beach is not subjected to wave activity except during storm conditions. It is actually in the supra-tidal zone—i.e., the zone above high tide where inundation by water is caused not by regular astronomical tides but rather by storm-generated tides. During nonstorm conditions the back-beach is relatively inactive except for wind action, which may move sediment. In most cases, there is an onshore component to the wind, and sediment is carried from the back-beach landward, typically forming dunes. Any obstruction on the back-beach, such as vegetation, pieces of driftwood, fences, or even trash discarded by people, results in wind-blown sand accumulation.

There are variations in beach forms along the shore as well as in those perpendicular to the shore. Most common is the rhythmic topography that is seen along the foreshore. A close look at the shoreline along most beaches will show that it is not straight or gently curved but rather that it displays a regularly undulating surface much like a low-amplitude sine curve. This is seen both on the plan view of the shoreline and the topography of the foreshore. The spacing is regular along a given reach of coast, but it may vary from place to place or from time to time at a given place. At some locations, concentrations of gravel or shells may develop, forming beach cusps (more or less triangular deposits that point seaward) during some wave conditions.

Although there is a common trend to the beach profile, some variation exists both because of energy conditions and because of the material making up the beach. Generally speaking, a beach that is accumulating sediment and experiencing low-energy conditions tends to have a steep foreshore, whereas the same beach would have a relatively gentle foreshore during storm conditions when erosion is prevalent. The grain size of beach sediment also is an important factor in the slope of the foreshore. In general, the coarser the constituent grains, the steeper the foreshore. Examples include the gravel beaches of New England, as contrasted to the gently sloping sand beaches of the Texas coast.

Coastal Dunes

Immediately landward of the beach are commonly found large, linear accumulations of sand known as dunes. (For coverage of dunes in arid and semiarid regions, see sand dune.) They form as the wind carries sediment from the beach in a landward direction and deposits it wherever an obstruction hinders further transport. Sediment supply is the key limiting factor in dune development and is the primary reason why some coastal dunes, such as those on the west Florida peninsula, are quite small, whereas others in such areas as the Texas coast and the Florida panhandle have large dunes.

Small wind-shadow dunes, or coppice mounds, actually may form on the backshore of the beach. If sediment continues to be supplied and beach erosion does not destroy them, these small sand accumulations will become foredunes, the seaward-most line of coastal dunes. It is in this fashion that a coast progrades, or grows seaward. Many barrier-island or strand-plain coasts exhibit numerous, essentially parallel dune ridges testifying to this type of growth.

The sediment in dunes tends to be fine to medium sand that is quite well sorted. Shell debris or other material is uncommon unless it is the same size or mass as the dune sand. There are various types of vegetation that grow on the dune surface and stabilize it. These grasses and vines often can be seen on the backshore portion of beaches as well. Dunes lacking vegetation are usually active and exhibit various signs of sand mobility. Most widespread are the nearly ubiquitous ripples that cover the dune surface. Large lobes of sand or even an entire dune may also move as wind blows across the dune. This activity results in cross stratification of the dune in large sweeping patterns of wedge-shaped packages of sand.

River Delta

A river delta is a low-lying plain or landform that occurs at the mouth of a river near where it flows into an ocean or another larger body of water. Deltas' greatest importance to human activities, fish and wildlife lay in their characteristic highly fertile soil and dense, diverse vegetation.

In order to fully appreciate the role deltas play in our larger ecosystem, it is first important to understand rivers. Rivers are defined as bodies of fresh water generally flowing from high elevations toward an ocean, a lake or another river; sometimes, even back into the ground.

Most rivers begin at high elevations where snow, rain, and other precipitation run downhill into creeks and small streams. These small waterways flow ever farther downhill, eventually meeting to form rivers.

Rivers flow toward oceans or other larger bodies of water, oftentimes combining with other rivers. Deltas exist as the lowest part of these rivers. It is in these deltas where a river's flow slows and spreads out to create sediment-rich dry areas and bio diverse wetlands.

Formation of River Deltas

The formation of a river delta is a slow process. As rivers flow toward their outlets from higher elevations, they deposit mud, silt, sand, and gravel particles at the mouths where rivers and larger, more sedentary bodies of water meet.

Over time these particles (called sediment or alluvium) build up at the mouth, extending into the ocean or lake. As these areas continue to grow the water becomes shallower and eventually, landforms begin to rise above the surface of the water, typically elevating to just above sea level.

As rivers drop enough sediment to create these landforms or areas of raised elevation, the remaining flowing water with the most power sometimes cuts across the land, forming different branches called distributaries.

Once formed, deltas are typically made up of three parts: the upper delta plain, the lower delta plain, and the subaqueous delta.

The upper delta plain makes up the area nearest to land. It is usually the area with the least water and highest elevation.

The lower delta plain is the middle of the delta. It is a transition zone between the dry upper delta and the wet subaqueous delta.

The subaqueous delta is the portion of the delta closest to the sea or body of water into which the river flows. This area is usually past the shoreline and it is below water level.

Types of River Deltas

Despite the generally universal processes by which river deltas are formed and organized, it is important to note that the world's deltas vary dramatically in structure, composition, and size due to factors such as origin, climate, geology, and tidal processes. These external factors contribute to an impressive diversity of deltas around the world. A delta's characteristics are classified based upon the specific factors contributing to its river's deposition of sediment -- typically the river itself, waves or tides.

The main types of deltas are wave-dominated deltas, tide-dominated deltas, Gilbert deltas, inland deltas, and estuaries.

As its name would imply, a wave-dominated delta such as the Mississippi River Delta is created by wave erosion controlling where and how much river sediment remains in a delta once it has been dropped. These deltas are usually shaped like the Greek symbol, delta (Δ).

Tide-dominated deltas such as the Ganges River Delta are formed by tides. Such deltas are characterized by a dendritic structure (branched, like a tree) due to newly-formed distributaries during times of high water.

Gilbert deltas are steeper and formed by deposition of coarse material. While it is possible for them to form in ocean areas, their formations are more commonly seen in mountainous areas where mountain rivers deposit sediment into lakes.

Inland deltas are deltas formed in inland areas or valleys where rivers may divide into many branches and rejoin farther downstream. Inland deltas, also called inverted river deltas, normally form on former lake beds.

Finally, when a river is located near coasts characterized by large tidal variations, they do not always form a traditional delta. Tidal variation often results in estuaries or a river that meets the sea, such as Saint Lawrence River in Ontario, Quebec, and New York.

Humans and River Deltas

River deltas have been important to humans for thousands of years because of their extremely fertile soils. Major ancient civilizations grew along deltas such as those of the Nile and the Tigris-Euphrates rivers, with the inhabitants of these civilizations learning how to live with their natural flooding cycles.

Many people believe that the ancient Greek historian Herodotus first coined the term delta nearly 2,500 years ago as many deltas are shaped like the Greek delta (Δ) symbol.

Deltas remain important to humans even today as, among many other things, a source of sand and gravel. Used in highway, building and infrastructure construction, these highly valuable materials quite literally build our world.

Delta land is also important in agricultural use. Witness the Sacramento-San Joaquin Delta in California. One of the most agriculturally diverse and productive areas in the state, the region successfully supports numerous crops from kiwi to alfalfa to tangerines.

Biodiversity and Importance of River Deltas

In addition to these human uses, river deltas boast some of the most bio diverse systems on the planet. As such, it is essential that these unique and beautiful havens of biodiversity remain as healthy habitat for the many species of plants, animals, insects, and fish -- some rare, threatened or endangered -- which call them home.

In addition to their biodiversity, deltas and wetlands provide a buffer for hurricanes, as open land often stands to weaken the impact of storms as they travel toward larger, more populated areas. The Mississippi River Delta, for example, buffers the impact of potentially strong hurricanes in the Gulf of Mexico.

Peninsula

A peninsula is a piece of land that is almost entirely surrounded by water but is connected to the mainland on one side.

Peninsulas can be very small, sometimes only large enough for a single lighthouse, for instance. Lighthouses often sit on peninsulas near rocky coastlines to warn sailors that they are getting close to land.

Peninsulas can also be very large. Most of the U.S. state of Florida is a peninsula that separates the Gulf of Mexico and the Atlantic Ocean.

Peninsulas are found on every continent. In North America, the narrow peninsula of Baja California, in Mexico, separates the Pacific Ocean and the Sea of Cortez, also called the Gulf of California. In Europe, the nations of Portugal and Spain make up the Iberian Peninsula. The so-called Horn of Africa, which juts into the Arabian Sea on central Africa's east coast, is a huge peninsula. The nations of North Korea and South Korea make up the Korean Peninsula in eastern Asia. In Australia, the Cape York Peninsula is only 160 kilometers (99 miles) from the island of New Guinea. The Antarctic Peninsula seems to point to the tip of South America, several hundred kilometers (miles) away.

This peninsula juts into the Black Sea.

Meander

A meander is one of a series of regular sinuous curves, bends, loops, turns, or windings in the channel of a river, stream, or other watercourse. It is produced by a stream or river swinging from side to side as it flows across its floodplain or shifts its channel within a valley. A meander is produced by a stream or river as it erodes the sediments comprising an outer, concave bank (cut bank) and deposits this and other sediment downstream on an inner, convex bank which is typically a point bar. The result of sediments being eroded from the outside concave bank and their deposition on an inside convex bank is the formation of a sinuous course as a channel migrates back and forth across the down-valley axis of a floodplain. The zone within which a meandering stream shifts its channel across either its floodplain or valley floor from time to time is known as a meander belt. It typically ranges from 15 to 18 times the width of the channel. Over time, meanders migrate downstream, sometimes in such a short time as to create civil engineering problems for local municipalities attempting to maintain stable roads and bridges.

The degree of meandering of the channel of a river, stream, or other watercourse is measured by its sinuosity. The sinuosity of a watercourse is the ratio of the length of the channel to the straight line down-valley distance. Streams or rivers with a single channel and sinuosities of 1.5 or more are defined as meandering streams or rivers.

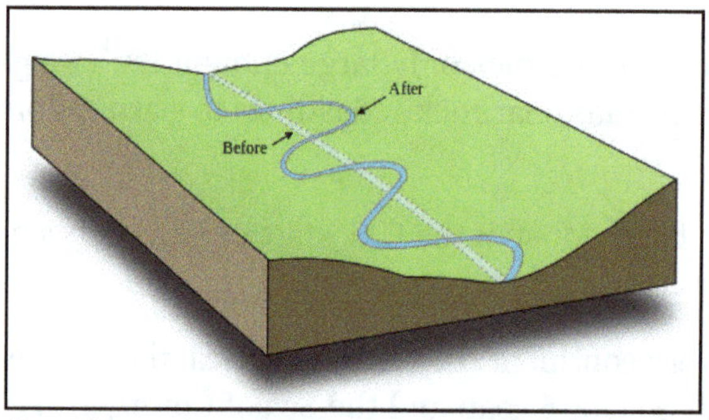

A streambed following a tilted valley. The maximum gradient is along the down-valley axis represented by a hypothetical straight channel. Meanders develop, which lengthen the course of the stream, decreasing the gradient.

Meanders of the *Rio Cauto* at Guamo Embarcadero, Cuba.

Meanders of the Vydrica river in Bratislava forest park, Slovakia.

Governing Physics

When a fluid is introduced to an initially straight channel which then bends, the sidewalls induce a pressure gradient that causes the fluid to alter course and follow the bend. From here, two opposing processes occur: (1) irrotational flow and (2) secondary flow. For a river to 'meander', secondary flow must dominate.

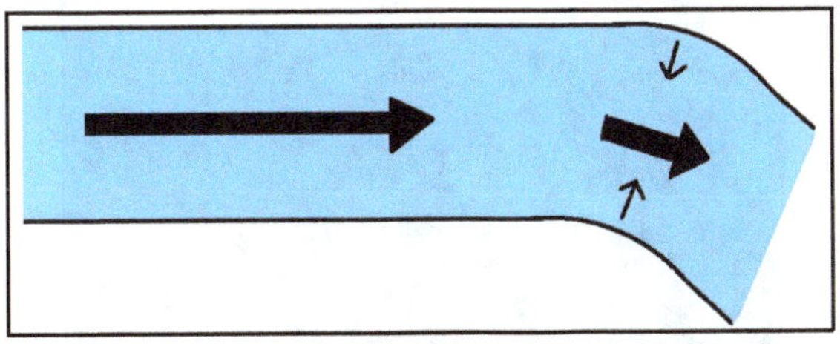

Straight channel culminating in a single bend.

Irrotational flow: From Bernoulli's equations, high pressure results in low velocity. Therefore, *in the absence of secondary flow* we would expect low fluid velocity at the outside bend and high fluid velocity at the inside bend. This classic fluid mechanics result is *irrotational vortex flow*. In the context of meandering rivers, its effects are dominated by those of secondary flow.

Secondary flow: A force balance exists between pressure forces pointing to the inside bend of the river and centrifugal forces pointing to the outside bend of the river. In the context of meandering rivers, a boundary layer exists within the thin layer of fluid that interacts with the river bed. Inside that layer and following standard boundary-layer theory, the velocity of the fluid is effectively zero. Centrifugal force, which depends on velocity, is also therefore effectively zero. Pressure force, however, remains unaffected by the boundary layer. Therefore, within the boundary layer, pressure force dominates and fluid moves along the bottom of the river from the outside bend to the inside bend. This initiates helicoidal flow: Along the river bed, fluid roughly follows the curve of the channel but is also forced toward the inside bend; away from the river bed, fluid also roughly follows the curve of the channel but is forced, to some extent, from the inside to the outside bend. Ultimately, the downstream velocity of the fluid is convectively transported to the outside bend, resulting in higher velocities at the outside bend. This secondary flow effect dominates over that of irrotational flow: In real meandering rivers, we observe higher downstream fluid velocities at the outside bends.

The higher (lower) velocities at the outside (inside) bend result in higher (lower) shear stresses and therefore results in erosion (deposition). Thus meander bends erode at the outside bend, causing the river to becoming increasingly sinuous (until cutoff events occur). Deposition at the inside bend occur such that for most natural meandering rivers, the river width remains nearly constant, even as the river evolves.

Meander Geometry

Uvac canyon meander, Serbia.

Meanders on the River Clyde, Scotland.

The technical description of a meandering watercourse is termed meander geometry or meander planform geometry. It is characterized as an irregular waveform. Ideal waveforms, such as a sine wave, are one line thick, but in the case of a stream the width must be taken into consideration. The bank full width is the distance across the bed at an average cross-section at the full-stream level, typically estimated by the line of lowest vegetation.

As a waveform the meandering stream follows the down-valley axis, a straight line fitted to the curve such that the sum of all the amplitudes measured from it is zero. This axis represents the overall direction of the stream.

At any cross-section the flow is following the sinuous axis, the centerline of the bed. Two consecutive crossing points of sinuous and down-valley axes define a meander loop. The meander is two consecutive loops pointing in opposite transverse directions. The distance of one meander along the down-valley axis is the meander length or wavelength. The maximum distance from the down-valley axis to the sinuous axis of a loop is the meander width or amplitude. The course at that point is the apex.

In contrast to sine waves, the loops of a meandering stream are more nearly circular. The curvature varies from a maximum at the apex to zero at a crossing point (straight line), also called an inflection, because the curvature changes direction in that vicinity. The radius of the loop is the straight line perpendicular to the down-valley axis intersecting the sinuous axis at the apex. As the loop is not ideal, additional information is needed to characterize it. The orientation angle is the angle between sinuous axis and down-valley axis at any point on the sinuous axis.

Concave bank and convex bank, Great Ouse Relief Channel, England.

A loop at the apex has an outer or concave bank and an inner or convex bank. The meander belt is defined by an average meander width measured from outer bank to outer bank instead of from centerline to centerline. If there is a flood plain, it extends beyond the meander belt. The meander is then said to be free—it can be found anywhere in the flood plain. If there is no flood plain, the meanders are fixed.

Various mathematical formulae relate the variables of the meander geometry. As it turns out some numerical parameters can be established, which appear in the formulae. The waveform depends ultimately on the characteristics of the flow but the parameters are independent of it and apparently are caused by geologic factors. In general the meander length is 10–14 times, with an average 11 times, the full bank channel width and 3 to 5 times, with an average of 4.7 times, the radius of curvature at the apex. This radius is 2–3 times the channel width.

Meander of the River Cuckmere in Southern England.

A meander has a depth pattern as well. The cross-overs are marked by riffles, or shallow beds, while at the apices are pools. In a pool direction of flow is downward, scouring the bed material. The major volume, however, flows more slowly on the inside of the bend where, due to decreased velocity, it deposits sediment.

The line of maximum depth, or channel, is the thalweg or thalweg line. It is typically designated the borderline when rivers are used as political borders. The thalweg hugs the outer banks and returns to center over the riffles. The meander arc length is the distance along the thalweg over one meander. The river length is the length along the centerline.

Formation

Meander formation is a result of natural factors and processes. The waveform configuration of a stream is constantly changing. Fluid flows around a bend in a vortex. Once a channel begins to follow a sinusoidal path, the amplitude and concavity of the loops increase dramatically due to the effect of helical flow sweeping dense eroded material towards the inside of the bend, and leaving the outside of the bend unprotected and therefore vulnerable to accelerated erosion, forming a positive feedback loop. In the words of Elizabeth A. Wood:

> 'this process of making meanders seems to be a self-intensifying process in which greater curvature results in more erosion of the bank, which results in greater curvature'

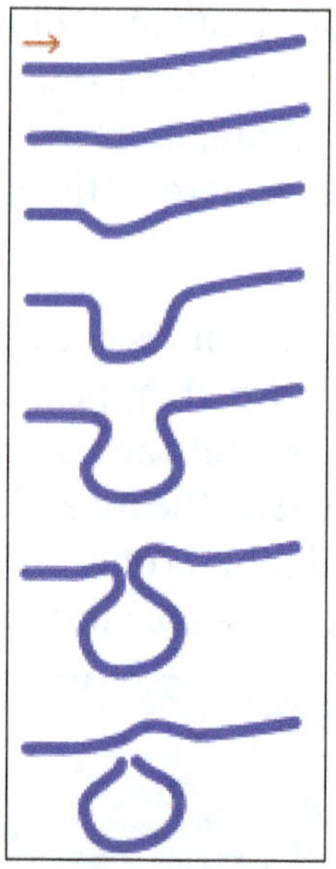

Life history of a meander.

The crosscurrent along the floor of the channel is part of the secondary flow and sweeps dense eroded material towards the inside of the bend. The cross-current then rises to the surface near the inside and flows towards the outside, forming the helical flow. The greater the curvature of the bend, and the faster the flow, the stronger is the crosscurrent and the sweeping.

Due to the conservation of angular momentum the speed on the inside of the bend is faster than on the outside.

Since the flow velocity is diminished, so is the centrifugal pressure. However, the pressure of the super-elevated column prevails, developing an unbalanced gradient that moves water back across the bottom from the outside to the inside. The flow is supplied by a counter-flow across the surface from the inside to the outside. This entire situation is very similar to the Tea leaf paradox. This secondary flow carries sediment from the outside of the bend to the inside making the river more meandering.

As to why streams of any size become sinuous in the first place, there are a number of theories, not necessarily mutually exclusive.

Stochastic Theory

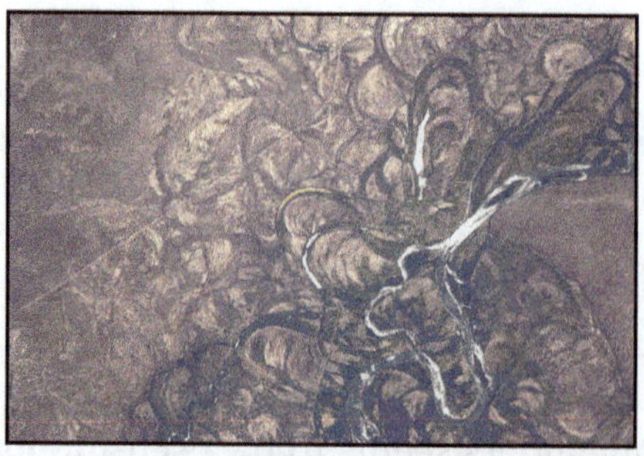

Meander scars, oxbow lakes and abandoned meanders in the broad flood plain of the Rio Negro, Argentina astronaut photo from ISS.

The stochastic theory can take many forms but one of the most general statements is that of Scheidegger: 'The meander train is assumed to be the result of the stochastic fluctuations of the direction of flow due to the random presence of direction-changing obstacles in the river path.' Given a flat, smooth, tilted artificial surface, rainfall runs off it in sheets, but even in that case adhesion of water to the surface and cohesion of drops produce rivulets at random. Natural surfaces are rough and erodible to different degrees. The result of all the physical factors acting at random is channels that are not straight, which then progressively become sinuous. Even channels that appear straight have a sinuous thalweg that leads eventually to a sinuous channel.

Equilibrium Theory

In the equilibrium theory, meanders decrease the stream gradient until an equilibrium between the erodibility of the terrain and the transport capacity of the stream is reached. A mass of water descending must give up potential energy, which, given the same velocity at the end of the drop as at the beginning, is removed by interaction with the material of the streambed. The shortest distance; that is, a straight channel, results in the highest energy per unit of length, disrupting the banks more, creating more sediment and aggrading the stream. The presence of meanders allows the stream to adjust the length to an equilibrium energy per unit length in which the stream carries away all the sediment that it produces.

Geomorphic and Morphotectonic Theory

Geomorphic refers to the surface structure of the terrain. Morphotectonic means having to do with the deeper, or tectonic (plate) structure of the rock. The features included under these categories are not random and guide streams into non-random paths. They are predictable obstacles that instigate meander formation by deflecting the stream. For example, the stream might be guided into a fault line (morphotectonic).

Associated Landforms

Cut Bank

A cut bank is an often-vertical bank or cliff that forms where the outside, concave bank of a meander cuts into the floodplain or valley wall of a river or stream. A cutbank is also known either as a river-cut cliff, river cliff, or a bluff and spelled as cutbank. Erosion that forms a cut bank occurs at the outside bank of a meander because helicoidal flow of water keeps the bank washed clean of loose sand, silt, and sediment and subjects it to constant erosion. As a result, the meander erodes and migrates in the direction of the outside bend, forming the cut bank.

As the cut bank is undermined by erosion, it commonly collapses as slumps into the river channel. The slumped sediment, having been broken up by slumping, is readily eroded and carried toward the middle of the channel. The sediment eroded from a cut bank tends to be deposited on the point bar of the next downstream meander, and not on the point bar opposite it. This can be seen in areas where trees grow on the banks of rivers; on the inside of meanders, trees, such as willows, are often far from the bank, whilst on the outside of the bend, the tree roots are often exposed and undercut, eventually leading the trees to fall into the river.

Meander Cutoff

The Rincon on Lake Powell in southern Utah. It is an incised cutoff (abandoned) meander.

A meander cutoff, also known as either a cutoff meander or abandoned meander, is a meander that has been abandoned by its stream after the formation of a neck cutoff. A lake that occupies an cutoff meander is known as an *oxbow lake*. Cutoff meanders that have cut downward into the underlying bedrock are known in general as incised cutoff meanders. As in the case of the Anderson Bottom Rincon, incised meanders that have either steep-sided, often vertical walls, are often, but not always, known as rincons in the southwest United States. *Rincon* in English is a nontechnical word in the southwest United States for either a small-secluded valley, an alcove or angular recess in a cliff, or a bend in a river.

Incised Meanders

Glen Canyon, US.

The meanders of a stream or river that has cut its bed down into the bedrock are known as either incised, intrenched, entrenched, inclosed or ingrown meanders. Some Earth scientists recognize and use a finer subdivision of incised meanders. Thornbury argues that *incised* or *inclosed meanders* are synonyms that are appropriate to describe any meander incised downward into bedrock and defines *enclosed* or *entrenched meanders* as a subtype of incised meanders (inclosed meanders) characterized by a symmetrical valley sides. He argues that the symmetrical valley sides are the direct result of rapid down cutting of a watercourse into bedrock. In addition, as proposed by Rich, Thornbury argues that incised valleys with a pronounced asymmetry of cross section, which he called *ingrown meanders*, are the result of the lateral migration and incision of a meander during a period of slower channel down cutting. Regardless, the formation of both entrenched meanders and ingrown meanders is thought to require that base level falls as a result of either relative change in mean sea level, isostatic or tectonic uplift, the breach of an ice or landslide dam, or regional tilting. Classic examples of incised meanders are associated with rivers in the Colorado Plateau, the Kentucky River Palisades in central Kentucky, and streams in the Ozark Plateau.

Goosenecks of the San Juan River, SE Utah.

It was initially either argued or presumed that an incised meander is characteristic of an antecedent stream or river that had incised its channel into underlying strata. An antecedent stream or river is one that maintains its original course and pattern during incision despite the changes in underlying rock topography and rock types. However, later geologists argue that the shape of an incised meander is not always, if ever, "inherited," e.g., strictly from an antecedent meandering stream where it meander pattern could freely develop on a level floodplain. Instead, they argue that as fluvial incision of bedrock proceeds, the stream course is significantly modified by variations in rock type and fractures, faults, and other geological structures into either *lithologically conditioned meanders* or *structurally controlled meanders*.

Oxbow Lakes

The oxbow lake, which is the most common type of fluvial lake, is a crescent-shaped lake that derives its name from its distinctive curved shape. Oxbow lakes are also known as cutoff lakes. Such lakes form regularly in undisturbed floodplains as a result of the normal process of fluvial meandering. Either a river or stream forms a sinuous channel as the outer side of its bends are eroded away and sediments accumulate on the inner side, which forms a meandering horseshoe-shaped bend. Eventually as the result of its meandering, the fluvial channel cuts through the narrow neck of the meander and forms a cutoff meander. The final break-through of the neck, which is called a neck cutoff, often occurs during a major flood because that is when the watercourse is out of its banks and can flow directly across the neck and erode it with the full force of the flood.

After a cutoff meander is formed, river water flows into its end from the river builds small delta-like feature into either end of it during floods. These delta-like features block either end of the cutoff meander to form a stagnant oxbow lake that is separated from the flow of the fluvial channel and independent of the river. During floods, the flood waters deposit fine-grained sediment into the oxbow lake. As a result, oxbow lakes tend to become filled in with fine-grained, organic-rich sediments over time.

Point Bar

A point bar, which is also known as a meander bar, is a fluvial bar that is formed by the slow, often episodic, addition of individual accretions of noncohesive sediment on the inside bank of a meander by the accompanying migration of the channel toward its outer bank. This process is called lateral accretion. Lateral accretion occurs mostly during high water or floods when the point bar is submerged. Typically, the sediment consists of either sand, gravel, or a combination of both. The sediment comprising some point bars might grade downstream into silty sediments. Because of the decreasing velocity and strength of current from the thalweg of the channel to the upper surface of point bar when the sediment is deposited the vertical sequence of sediments comprising a point bar becomes finer upward within an individual point bar. For example, it is typical for point bars to fine upward from gravel at the base to fine sands at the top. The source of the sediment is typically upstream cut banks from which sand, rocks and debris has been eroded, swept, and rolled across the bed of the river and downstream to the inside bank of a river bend. On the inside bend, this sediment and debris is eventually deposited on the slip-off slope of a point bar.

Scroll-bars

Scroll-bars are a result of continuous lateral migration of a meander loop that creates an asymmetrical ridge and swale topography on the inside of the bends. The topography is generally parallel to the meander, and is related to migrating bar forms and back bar chutes, which carve sediment from the outside of the curve and deposit sediment in the slower flowing water on the inside of the loop, in a process called lateral accretion. Scroll-bar sediments are characterized by cross bedding and a pattern of fining upward. These characteristics are a result of the dynamic river system, where larger grains are transported during high energy flood events and then gradually die down, depositing smaller material with time. Deposits for meandering rivers are generally homogeneous and laterally extensive unlike the more heterogeneous braided river deposits. There are two distinct patterns of scroll-bar depositions; the eddy accretion scroll bar pattern and the point-bar scroll pattern. When looking down the river valley they can be distinguished because the point-bar scroll patterns are convex and the eddy accretion scroll bar patterns are concave.

Scroll bars often look lighter at the tops of the ridges and darker in the swales. This is because the tops can be shaped by wind, either adding fine grains or by keeping the area unvegetated, while the darkness in the swales can be attributed to silts and clays washing in during high water periods. This added sediment in addition to water that catches in the swales is in turn is a favorable environment for vegetation that will also accumulate in the swales.

Slip-off Slope

Depending upon whether a meander is part of an entrenched river or part of a freely meandering river within a floodplain, the term slip-off slope can refer to two different fluvial landforms that comprise the inner, convex, bank of a meander loop. In case of a freely meandering river on a floodplain, a *slip-off slope* is the inside, gently sloping bank of a meander on which sediments episodically accumulate to form a point bar as a river meanders. This type of slip-off slope is located opposite the cutbank. This term can also be applied to the inside, sloping bank of a meandering tidal channel.

In case of an entrenched river, a *slip-off slope* is a gently sloping bedrock surface that rises from the inside, concave bank of an asymmetrically entrenched river. This type of slip-off slope is often covered by a thin, discontinuous layer of alluvium. It is produced by the gradual outward migration of the meander as a river cuts downward into bedrock. A terrace on the slip-off slope of a meander spur, known as slip-off slope terrace, can formed by a brief halt during the irregular incision by an actively meandering river.

Derived Quantities

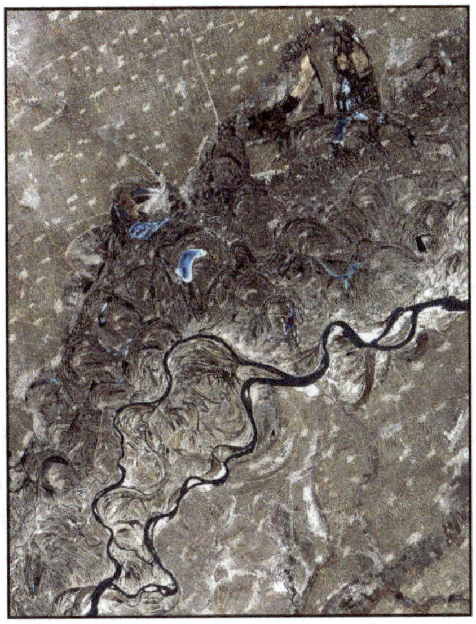

Meanders, scroll-bars and oxbow lakes in the Songhua River.

The meander ratio or sinuosity index is a means of quantifying how much a river or stream meanders (how much its course deviates from the shortest possible path). It is calculated as the length of the stream divided by the length of the valley. A perfectly straight river would have a meander ratio of 1 (it would be the same length as its valley), while the higher this ratio is above 1, the more the river meanders.

Sinuosity indices are calculated from the map or from an aerial photograph measured over a distance called the reach, which should be at least 20 times the average full bank channel width. The length of the stream is measured by channel, or thalweg, length over the reach, while the bottom value of the ratio is the down valley length or air distance of the stream between two points on it defining the reach.

The sinuosity index plays a part in mathematical descriptions of streams. The index may require elaboration, because the valley may meander as well—i.e., the down valley length is not identical to the reach. In that case the valley index is the meander ratio of the valley while the channel index is the meander ratio of the channel. The channel sinuosity index is the channel length divided by the valley length and the standard sinuosity index is the channel index divided by the valley index. Distinctions may become even more subtle.

Sinuosity Index has a non-mathematical utility as well. Streams can be placed in categories arranged by it; for example, when the index is between 1 and 1.5 the river is sinuous, but if between 1.5 and 4, then meandering. The index is a measure also of stream velocity and sediment load, those quantities being maximized at an index of 1 (straight).

Cliffed Coast

Coastal cliffs are very steep rock faces near the sea that are greater than 5 m in height. They may ascend in steps and have ledges, crevices and overhangs. Coastal cliffs may rise directly from the sea or be separated from it by a narrow shore. Basic rocks include basalt, andesite, diorite, gabbro, and tuffaceous mudstones and sandstones. Cliffs and outcrops provide many varied habitats: from bare rock colonized only by mosses and lichens, to deeper soils supporting woody vegetation; from highly exposed situations, to heavily shaded and sheltered habitats; and from very dry to permanently wet surfaces. Coastal cliffs are particularly influenced by salt spray, with halophytes and succulents.

Cliffed Coast Morphology

Cliff Form and Occurrence

While cliffed coasts are occasionally formed in cohesionless sands where plant roots and soil moisture provide some strength, most cliffed shorelines develop in material that possesses strength due to cohesion provided by the bonding of clay minerals, cementation by chemical precipitates or the crystal bonding of igneous and metamorphic rocks. The term cliff is used here for all shorelines with a steep sub aerial slope –strictly speaking some portion of the slope should exceed 40°, the height of the cliff should exceed the maximum height of wave run-up and overtopping (though wave spray may reach the top of the cliff). If the cliff is so low that wave overtopping can occur then the shoreline feature is termed a bank. The term bluff can be used interchangeably with cliff, but here its use is restricted to describing cliffs formed in unconsolidated or not consolidated sediments, including sand, silt, clay and till. These cliffed coasts are termed cohesive coasts to distinguish them from coasts formed in much stronger bedrock where coastal processes modify the shoreline relatively slowly. While the terminology is a bit ambiguous because muddy coasts (e.g., estuarine mudflats) are also cohesive and because cohesion can equally apply to crystalline rocks, practically it is useful to make the distinction between them.

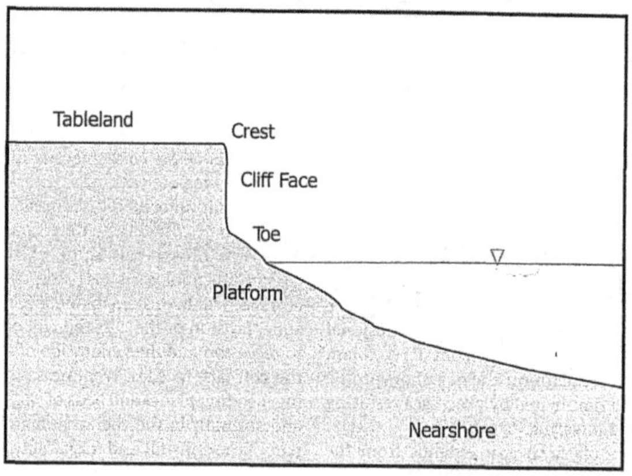

The presence of a cliffed shoreline reflects the existence of relatively high ground near the coast and this in turn may reflect tectonic forces on a continental scale, local folding and faulting, fluvial or glacial erosion, water level change, or simply recession due to coastal erosion on a gently sloping plain. Thus, the height of the coastal cliff is controlled primarily by the relief of the coastal zone and secondarily due to the operation of coastal processes.

In examining the processes and components that make up the coastal cliff recession system it is useful first of all to describe the morphological components of the system in a profile normal to the shoreline (figure) Not all features will be present on all cliffed coasts: The major components are:

a) The tableland or area inland from the cliff top.

b) The cliff top that marks the change in slope from the tableland to the cliff face and is the transition zone to the area that slopes down to the water.

c) The sub aerial difffacethat extends from the cliff top to the toe of the slope where it intersects the beach or platform. This area is dominated by erosion due to processes resulting from mass wasting as well as overland flow and gullying. In the case of plunging cliffs the face extends below the water level.

d) The cliffs toe that is the transition area between the sub aerial cliff and the beach and shore platform. The upper limit of the cliff toe is marked by height to which wave action (not including spray) can reach and the lower limit by junction with the more gently sloping shore platform. It also marks the transition from the cliff face, which is dominated by sub aerial erosional processes leading to horizontal recession, and the shore platform and nearshore profile, which are dominated by processes resulting in vertical lowering.

e) The shore platform that extends from the base of the cliff offshore to a point at, or just below spring low tide. The shore platform may be overlain 'by varying amounts of surficial sediments. The platform itself is subject to wave action as well as weathering processes during sub aerial exposure.

f) The nearshore slope that forms the subaqueous extension of the intertidal platform, and is a zone of shoaling and breald.ng waves extending offshore to the limit of wave erosion and transport of sediment.

Cliff Coast Classification

Two major types of cliffed coast can be recognised on the basis of the profile form normal to the shore and sea level (Figure below). Plunging cliffs occur where the cliff extends below the water line to some considerable depth (Figures below). Waves break directly against the cliff face, and there is no beach, ramp or platform that would lead to wave breald.ng offshore and to the accumulation of sediments. Plunging cliffs usually occur in resistant bedrock where the slope and relief are determined by tectonic events (e.g., folding or block faulting) or where erosion by glaciers or rivers, and subsequent drowning due to sea level rise, has produced a steep cliff with deep water at the cliff base. Over time, jointing and vertical weal messes may be exploited by wave action to produce chimneys and small pocket beaches, and irregularities in the cliff face may be enhanced by runoff and gully development. However, erosional processes are generally very slow on these coasts so these features form slowly. Plunging cliffs are not generally found on cohesive coasts because the cliff material is too weal< to withstand direct wave attack for very long and cliff recession soon leads to the formation of a sloping platform and beach.

Where the overall slope of the inherited coastal morphology is less steep the toe of the cliff face will be located in or above the intertidal zone. Erosion of the cliff toe will occur, leading to recession and the generation of a platform as the cliff face recedes. Weal messes such as jointing, bedding planes and beds of varying lithology and strength in the rock malting up the lower part of the cliff will lead to spatially uneven rates of erosion and the development of a variety of erosional forms such as notches, blowholes, caves, arches and stacks. The coastline tends to become highly irregular and the inner nearshore is often rocky with a variety of shallow reefs and emergent boulders (figures below). These features are absent on cohesive coasts where rapid erosion on the beach and shallow nearshore quickly removes any irregularities.

Horizontal erosion is focused at the toe of the cliff, and recession of the cliff itself will tend to produce a quasi-horizontal erosion surface or platform. However, erosional processes in the intertidal and sub-tidal zones also act on the platform leading to vertical lowering of the surface. This in turn generates a profile with an intertidal zone that slopes away from the base of the cliff and grades into the underwater nearshore profile without any abrupt transition. Type A platforms are the most common form of platform on cliffed shorelines, particularly in rocks of moderate to low strength and in areas where sand and gravel are present in the intertidal zone. However, in some areas vertical erosion of the platform in the intertidal and shallow sub-tidal zone is relatively slow compared to horizontal recession of the cliff toe. This leads to the development of a Type B shore platform that has a nearly horizontal surface away from the base of the cliff, and then terminate abruptly in a seaward drop to the nearshore (Figures below). The elevation of the platform may be close to the high-tide level, the low tide level, or somewhere in between. Type A platforms reflect conditions where vertical lowering of the platform in the intertidal area is similar to that in the inner nearshore and keeps pace with horizontal retreat of the cliff toe. The development of the nearly horizontal platform associated with type B must therefore reflect much slower vertical lowering of the intertidal platform compared to the recession of the cliff toe. The quasi-horizontal Type B platforms have generated much interest and there is considerable debate about the processes operating on them and the controls on their origin. We will examine this problem at the end of the chapter after we look at cliffed coast processes horizontal platform associated with type B must therefore reflect much slower vertical lowering of the intertidal platform compared to the recession of the cliff toe. The quasi-horizontal Type B platforms have generated much interest and there is considerable debate about the processes operating on them and the controls on their origin. We will examine this problem at the end of the chapter after we look at cliffed coast processes generally, and at processes operating on cohesive and rock coasts.

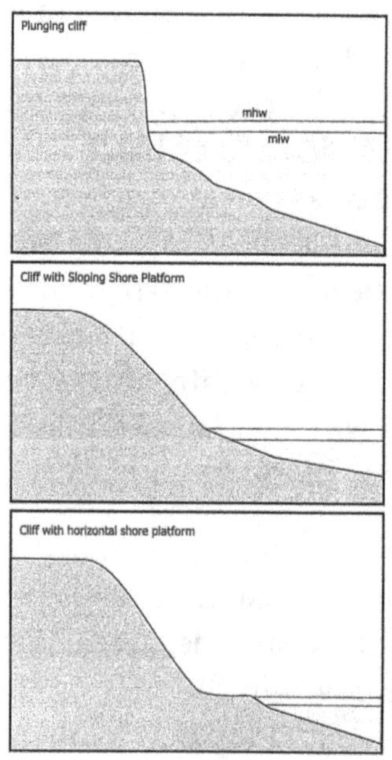

Profiles associated with the major types of cliffed coast.

Volcanic and Slope Landforms

Volcanic landforms are controlled by the geological processes that form them and act on them after they have formed. Thus, a given volcanic landform will be characteristic of the types of material it is made of, which in turn depends on the prior eruptive behavior of the volcano. Although later processes can modify the original landform, we should be able to find clues in the modified form that lead us to conclusions about the original formation process.

Shield Volcanoes

- A shield volcano is characterized by gentle upper slopes (about 50°) and somewhat steeper lower slopes (about 10°).

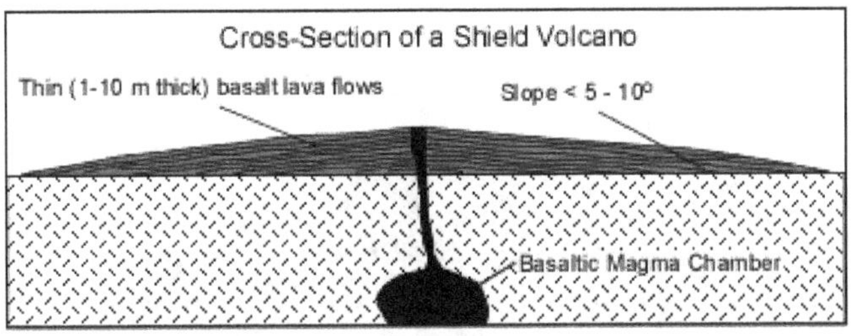

- Shield volcanoes are composed almost entirely of relatively thin lava flows built up over a central vent.
- Most shields were formed by low viscosity basaltic magma that flows easily down slope away form the summit vent.
- The low viscosity of the magma allows the lava to travel down slope on a gentle slope, but as it cools and its viscosity increases, its thickness builds up on the lower slopes giving a somewhat steeper lower slope.
- Most shield volcanoes have a roughly circular or oval shape in map view.
- Very little pyroclastic material is found within a shield volcano, except near the eruptive vents, where small amounts of pyroclastic material accumulate as a result of fire fountaining events.
- Shield volcanoes thus form by relatively non-explosive eruptions of low viscosity basaltic magma.

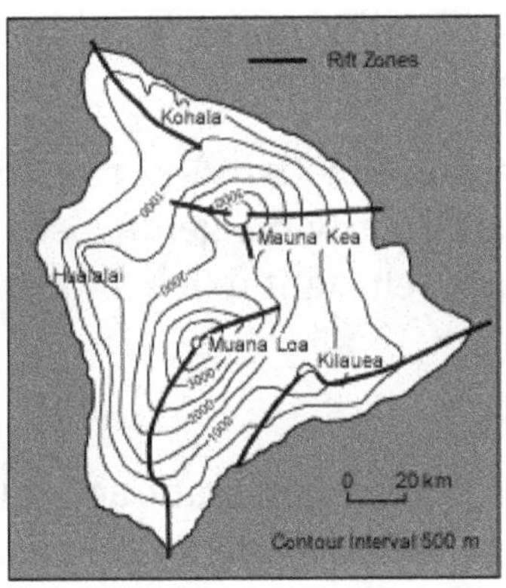

- Vents for most shield volcanoes are central vents, which are circular vents near the summit. Hawaiian shield volcanoes also have flank vents, which radiate from the summit and take the form of en-echelon fractures or fissures, called rift zones, from which lava flows are emitted. This gives Hawaiian shield volcanoes like Kilauea and Mauna Loa their characteristic oval shape in map view.

Stratovolcanoes

- Have steeper slopes than shield volcanoes, with slopes of 6 to 100 low on the flanks to 30° near the top.
- The steep slope near the summit is due partly to thick, short viscous lava flows that do not travel far down slope from the vent.

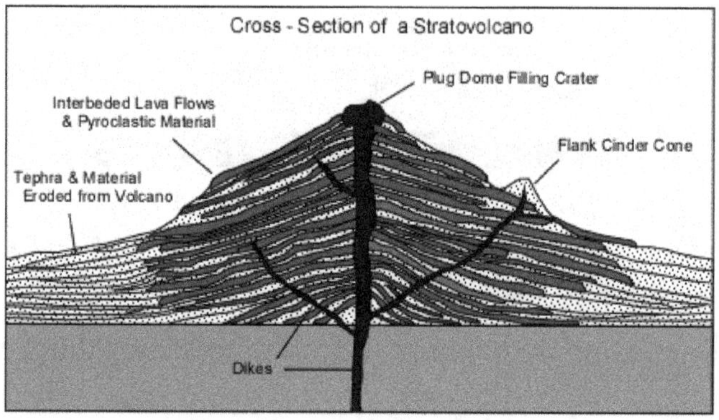

- The gentler slopes near the base are due to accumulations of material eroded from the volcano and to the accumulation of pyroclastic material.
- Stratovolcanoes show inter-layering of lava flows and pyroclastic material, which is why they are sometimes called composite volcanoes. Pyroclastic material can make up over 50% of the volume of a stratovolcano.
- Lavas and pyroclastics are usually andesitic to rhyolitic in composition.
- Due to the higher viscosity of magmas erupted from these volcanoes; they are usually more explosive than shield volcanoes.
- Stratovolcanoes sometimes have a crater at the summit that is formed by explosive ejection of material from a central vent. Sometimes the craters have been filled in by lava flows or lava domes, sometimes they are filled with glacial ice, and less commonly they are filled with water.
- Long periods of repose (times of inactivity) lasting for hundreds to thousands of years, make this type of volcano particularly dangerous, since many times they have shown no historic activity, and people are reluctant to heed warnings about possible eruptions.

Maars

- Maars result from phreatic or phreatomagmatic activity, wherein magma heats up groundwater, pressure builds as the water to turns to steam, and then the water and preexisting rock (and some new magma if the eruption is phreatomagmatic) are blasted out of the ground to form a tephra cone with gentle slopes.

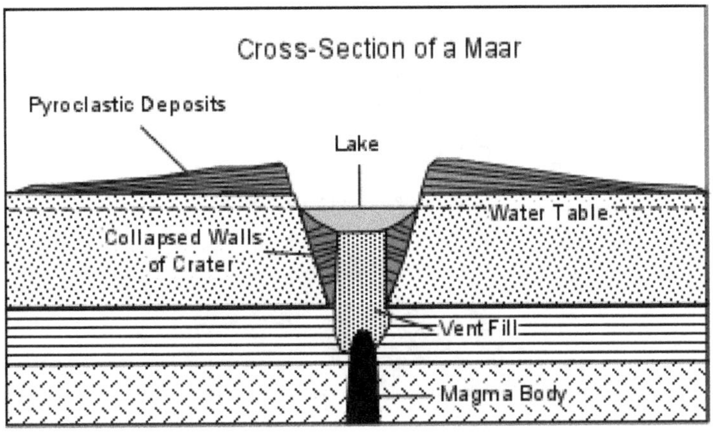

Parts of the crater walls eventually collapse back into the crater, the vent is filled with loose material, and, if the crater still is deeper than the water table, the crater fills with water to form a lake, the lake level coinciding with the water table.

Volcanic Domes

- Volcanic Domes result from the extrusion of highly viscous, gas poor andesitic and rhyolitic lava. Since the viscosity is so high, the lava does not flow away from the vent, but instead piles up over the vent.
- Blocks of nearly solid lava break off the outer surface of the dome and roll down its flanks to form a breccia around the margins of domes.
- The surface of volcanic domes are generally very rough, with numerous spines that have been pushed up by the magma from below.

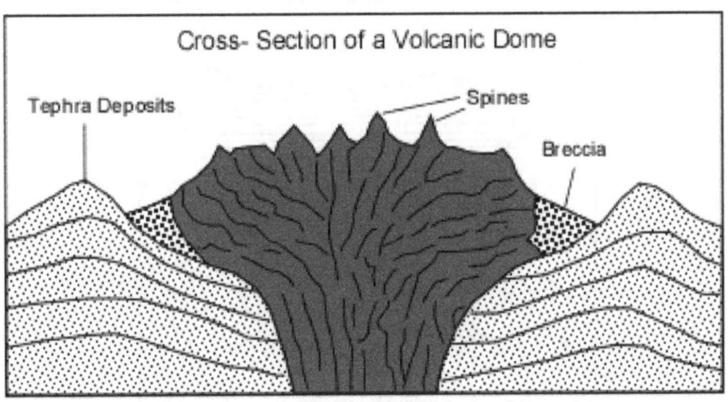

- Most dome eruptions are preceded by explosive eruptions of more gas rich magma, producing a tephra cone into which the dome is extruded.
- Volcanic domes can be extremely dangerous. Because they form unstable slopes that may collapse to expose gas-rich viscous magma to atmospheric pressure. This can result in lateral blasts or Pelean type pyroclastic flow eruptions.

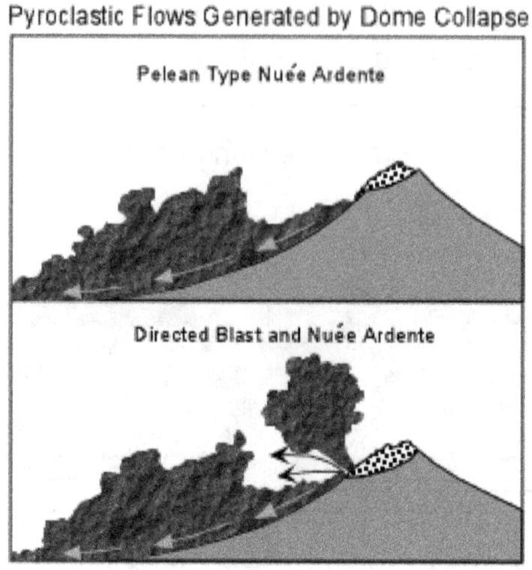

Resurgent Domes

- After the formation of a caldera by collapse, magma is sometimes re-injected into the area below the caldera. This can result in uplift of one or more areas within the caldera to form a resurgent dome. Two such resurgent domes formed in the Yellowstone caldera.
- If magma leaks back to the surface during this resurgent doming, then eruptions of small volcanic domes can occur in the area of the resurgent domes.

Geysers, Fumaroles and Hot Springs

- A fumarole is vent where gases, either from a magma body at depth, or steam from heated groundwater, emerges at the surface of the Earth. Since most magmatic gas is H_2O vapor, and since heated groundwater will produce H_2O vapor, fumaroles will only be visible if the water condenses. (H_2O vapor is invisible, unless droplets of liquid water have condensed).

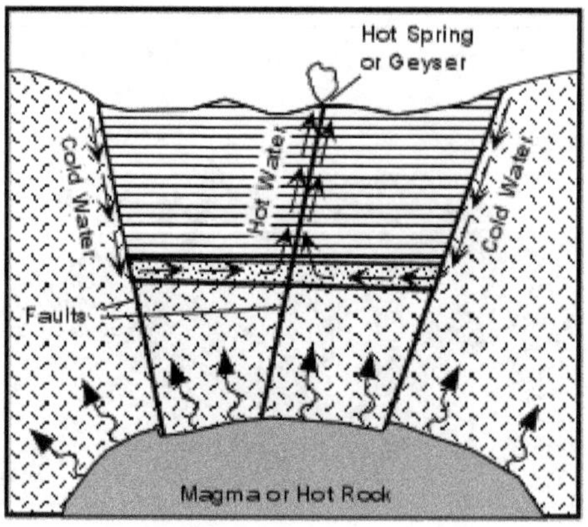

After Williams and McBirney, 1985

- Hot springs or thermal springs are areas where hot water comes to the surface of the Earth. Cool groundwater moves downward and is heated by a body of magma or hot rock. A hot spring results if this hot water can find its way back to the surface, usually along fault zones.

Minerals dissolved in the high temperature water are often precipitated when the water cools at the surface. This produces spectacular deposits of travertine (chemically precipitated calcite, or siliceous sinter.

Bacteria forming microbial mats under the water are responsible for the coloration often seen in hot springs. Different species, with different colors thrive at different temperatures.

- A geyser results if the hot spring has a plumbing system that allows for the accumulation of steam from the boiling water. When the steam pressure builds so that it is higher than the pressure of the overlying water in the system, the steam will move rapidly toward the surface, causing the eruption of the overlying water. Some geysers, like Old Faithful in Yellowstone Park, erupt at regular intervals. The time between eruptions is controlled by the time it takes for the steam pressure to build in the underlying plumbing system.

Volcanic Cone

A volcanic cone is a triangle-shaped hill formed as material from volcanic eruptions piles up around the volcanic vent, or opening in Earth's crust.

Most volcanic cones have one volcanic crater, or central depression, at the top. They are probably the most familiar type of volcanic mountain.

Major Types of Volcanic Cones

Composite Cones

Composite cones are some of the most easily recognizable and imposing volcanic mountains, with sloping peaks rising several thousand meters above the landscape.

Also known as stratocones, composite cones are made up of layers of lava, volcanic ash, and fragmented rocks. These layers are built up over time as the volcano erupts through a vent or group of vents at the summit's crater. The eruptions that form these cones, called Plinian eruptions, are violently explosive and often dangerous.

One of the most famous stratocones in the world is Mount Fuji, Japan. The tallest mountain in Japan, Mount Fuji towers 3,776 meters (12,380 feet) above the surrounding landscape. Mount Fuji last erupted in 1707, but is still considered an active volcano.

Mount Rainier, Washington, is another stratocone. Mount Rainier rises 4,392 meters (14,410 feet) above sea level. Over the past half million years, Mt. Rainier has produced a series of alternating lava eruptions and debris eruptions. These eruptions have given Mt. Rainier the classic layered structure and sloping shape of a composite cone. Unlike Mount Fuji, Mount Rainier's composite cone has been carved down by a series of glaciers, giving it a craggy and rugged shape.

Cinder Cones

Cinder cones, sometimes called scoria cones or pyroclastic cones, are the most common types of volcanic cones. They form after violent eruptions blow lava fragments into the air, which then solidify and fall as cinders around the volcanic vent. Usually the size of gravel, these cinders are filled with many tiny bubbles trapped in the lava as it solidifies. Cinder cones stand at heights of tens of meters to hundreds of meters.

Cinder cones may form by themselves or when new vents open on larger, existing volcanoes. Mauna Kea, a volcano on the American island of Hawaii, and Mount Etna, a volcano on the Italian island of Sicily, are both covered with hundreds of cinder cones.

Other Types of Volcanic Cones

Spatter Cones

Volcanoes often eject small amounts of gaseous lava blobs into the air. These lava blobs, called spatter, are heavy and viscous. Viscosity refers to a substance's resistance to flow. In this case, it refers to the spatter's thickness. The viscosity of spatter means it often does not have time to cool before hitting the ground.

The lava blobs in spatter stick together as they land, piling up to form steep-sided spatter cones. Most spatter cones are very small, ranging between 1 and 5 meters (3 to 16 feet) in height, because they result from minor volcanic activity. They often form in linear groups along an eruptive fissure, or long crack, on the flank of an active volcano. A small spatter cone is called a hornito.

Spatter cones can be found in and around the Pu'u 'Ō'ō region of Mount Kilauea in Hawaii. Continuously erupting since 1983, Kilauea's volcanic activity is characterized by the fountaining of hot lava, making it the perfect incubator for spatter cones.

Tuff Cones

Unlike spatter cones that form from lava fountaining, tuff cones form from the interaction between rising magma and bodies of water. Tuff cones are sometimes called ash cones.

When heated rapidly by lava, water flashes to steam and expands violently, fragmenting huge amounts of lava into plumes of very fine grains of ash. This ash falls around the volcanic vent, creating an ash cone. Over time, the ash weathers into a rock known as tuff.

Tuff cones have steep sides and often stand between 100 and 300 meters (328 to 984 feet) high. They are much wider and have broader craters than spatter cones because they result from shallow explosions that eject materials sideways rather than upwards.

Diamond Head, the famous volcano near Honolulu, Hawaii, is an enormous tuff cone. The mountain is the result of a brief volcanic eruption about 200,000 years ago. During Diamond Head's eruption period, the mountain rose from the ocean, and lava interacted with water and even a nearby coral reef. Today, Diamond Head's rim is about a kilometer (.62 mile) from the coast, and rises about 232 meters (760 feet) above sea level.

Wizard Island is a small cinder cone located inside the remains of a much larger volcano, Mount Mazama. Mount Mazama was a composite cone volcano that collapsed about 7,700 years ago and formed Crater Lake, Oregon.

Volcanic Crater

A volcanic crater is a roughly circular depression in the ground caused by volcanic activity. It is typically a bowl-shaped feature within which occurs a vent or vents. During volcanic eruptions, molten magma and volcanic gases rise from an underground magma chamber, through a tube-shaped conduit, until they reach the crater's vent, from where the gases escape into the atmosphere and the magma is erupted as lava. A volcanic crater can be of large dimensions, and sometimes of great depth. During certain types of explosive eruptions, a volcano's magma chamber may empty enough for an area above it to subside, forming a type of larger depression known as a caldera.

Craters on Mount Cameroon.

Geomorphology

In most volcanoes, the crater is situated at the top of a mountain formed from the erupted volcanic deposits such as lava flows and tephra. Volcanoes that terminate in such a summit crater are usually of a conical form. Other volcanic craters may be found on the flanks of volcanoes, and these are commonly referred to as flank craters. Some volcanic craters may fill either fully or partially with rain and/or melted snow, forming a crater lake.

A crater may be breached during an eruption, either by explosions or by lava, or through later erosion. Breached craters have a much lower rim on one side.

Some volcanoes, such as maars, consist of a crater alone, with scarcely any mountain at all. These volcanic explosion craters are formed when magma rises through water-saturated rocks, which causes a phreatic eruption. Volcanic craters from phreatic eruptions often occur on plains away from other obvious volcanoes. Not all volcanoes form craters.

The volcanic crater of a Tangkuban Parahu mount, Bandung, Indonesia.

Plateau

Plateaus can be formed by the erosion of the surrounding areas by rivers, flooding, and glacier. Repeated lava flows can also form a plateau over a period of time. Plateau is formed when the magma pushes up towards the surface of the earth's crust. There are many ways in which plateaus can develop. The magma does not break through but instead raises a section of the crust up as it raises and creates a plateau. It is also formed by the effects of that millions of years of wind and water erosion can have on land areas. Water that flows in specific pattern layers the rock at time and sculptures the channels that will continuously dig deeper. As the water digs its way down, steep slopes are created and that forms the edges of the plateau. Plateaus are very useful because they are rich in mineral deposits. As a result, many of the mining areas in the world are located in the plateau areas.

Types of the Plateau

- Plateaus are divided into three types.
- It is based on location, where they are situated.
  1. Intermontane plateaus
  2. Piedmont plateaus
  3. Continental plateaus

Volcanic Plateau

- It is formed by many small volcanic eruptions which slowly build up by time which forms plateau resulting in lava flows. Example- Deccan traps.

Tibetan Plateau

- It is called the roof of the world because it is the highest and largest plateau in the world.
- It is formed due to the collision of the Indo-Australian and Eurasian tectonic plates.
- The Himalayan range of the south mountain is surrounded by it.

Columbia –Snake Plateau

- It is formed by the volcanic eruptions with a related coating of basalt lava.
- It meets the river Columbia and its tributary snake meets in this plateau.

Deccan Plateau

- It is the plateau which forms the southern part of India and is the largest plateau.

- The Western Ghats and the Eastern Ghats are bordered by Deccan plateau.
- This is also known for containing some unique fossils.

Katanga Plateau

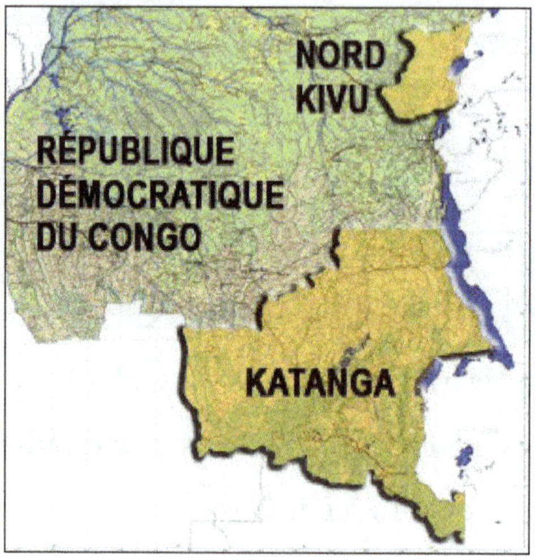

- Katanga plateau is famous for copper production.
- This lies in Congo.
- Minerals found here is zinc, silver, gold, and cobalt are mined here.

Facts about Plateau

- Plateaus can be formed by a number of processes like upwelling of volcanic magma, extrusion of lava, and erosion by water and glaciers.
- Plateaus can also be built up lava spreading outward from cracks and weak areas in the crust.
- Plateaus can also be formed by the erosional processes of glaciers on mountain ranges, leaving them sitting between the mountain ranges.
- Piedmont plateaus are bordered on one side by mountains and on the other by plain or sea

Plains

A plain is a type of landform made up of a flat area that can exist in valleys, lowlands, on plateaus, or uplands. They are formed by a variety of weather and geological phenomena including water deposits, ice, wind, erosion, and even lava. Plains make excellent agricultural ground in many places because of their rich soil and their relatively flat landscape. There are different types of plains including structural plains, erosional plains, and depositional plains, as well as coastal plains and flood plains. Natural plant life on plains is dependent on the climate and can range from thick forests, to grasslands, and everything in between.

Interesting Plains Facts

- Structural plains tend to be large flat surfaces that make up extensive lowlands.
- Erosional plains are those that have been created by erosion die to glaciers, wind, running water and rivers.
- Depositional plains are created when material is deposited from rivers, glaciers, waves and wind. Sometimes these plains are very fertile because of the type of material that has been deposited there.
- Depositional plains are classified as alluvial plains, or glacial plains.
- Alluvial plains are created by a river that deposits material that becomes the soil.
- Flood plains are plains that experience periodic flooding or just occasional flooding. One of the most famous flood plains is the one surrounding the Nile River in Africa.
- A lacustrine plain is a plain that was originally the bottom of a lake.
- A lava plain is formed when lava creates sheets over time, eventually becoming rich soil.
- Glacial plains are formed when a glacier moves across land and the force of gravity creates the large flat surface.
- An abyssal plain is an area of the ocean basin that is either flat or sloping very gently.
- Plains that exist on other planets are referred to as planitia (Latin for the word plain). Mars has Hellas Planitia and Venus has Sedna Planitia.
- The Great Plains in North America cover areas of ten states including New Mexico, Texas, Oklahoma, Colorado, Kansas, Wyoming, Nebraska, South Dakota, North Dakota, and Montana. Weather on the Great Plains can be extreme at times, which has led to a lot of irrigation to help compensate for drought conditions that can occur.
- Coastal plains are lowlands that stretch along the shore and slope towards it. An example of a coastal plain is the Atlantic Coastal Plain. This plain stretches from Florida to Nova Scotia.
- Many of the world's rivers are surrounded by plains.
- Plains cover approximately one-third of the land on earth.

- Every continent on earth has plains in one form or another.
- In North America, where some plains are grasslands, these are referred to as prairies. These types of plains have warm summers and cold winters.
- Mexico has a forested plain called Tabasco Plain that is home to all sorts of vegetation including trees and shrubs.
- A tropical grassland plain is called a savannah. Savannahs are warm all year and tend to have scattered trees. Africa's main savannah is the Serengeti.
- In very cold climates like the Arctic, plains are often frozen. They are referred to as tundra.

References

- Getis, Arthur; Judith Getis and Jerome D. Fellmann (2000). Introduction to Geography, Seventh Edition. McGraw Hill. p. 99. ISBN 0-697-38506-X
- Aeolian-landforms-what-is-a-yardang: worldatlas.com, Retrieved 09, May 2020
- Neuendorf, K.E.K., J.P. Mehl Jr., and J.A. Jackson, 2005, Glossary of Geology. Springer-Verlag, New York City, 779 pp., ISBN 3-540-27951-2
- What-are-landforms-and-major-types-of-landforms-on-earth: eartheclipse.com, Retrieved 29, July 2020
- Mountain-and-glacial-landforms-what-is-a-cirque: worldatlas.com, Retrieved 15, March 2020
- Schmidt, E. D.; Semmel, A.; Frechen, M. (2011). "Luminescence dating of the loess/palaeosol sequence at the gravel quarry Gaul/Weilbach, Southern Hesse (Germany)". Quaternary Science Journal. 60 (1): 116–125 doi:10 3285/eg 60 1 08
- Geography-of-river-deltas-1435824: thoughtco.com, Retrieved 30,June 2020
- The-glacial-landforms-and-cycle-of-erosion-744: geographynotes.com, Retrieved 11,April 2020
- Murray, A. S.; Wintle, A. G. (2000). "Luminescence dating of quartz using an improved single aliquot regenerative-dose protocol". Radiation Measurements. 32: 57–73. Bibcode:2000RadM...32...57M. doi:10.1016/S1350-4487(99)00253-X
- Coastal-landform, science: britannica.com, Retrieved 27,May 2020
- Thiel, C.; Buylaert, J. P.; Murray, A. S.; Terhorst, B.; Tsukamoto, S.; Frechen, M.; Sprafke, T. (2011). "Investigating the chronostratigraphy of prominent palaeosols in Lower Austria using post-IR IRSL dating". Quaternary Science Journal. 60.(1): 137–152. doi:10.3285/eg.60.1.10
- Valley, science: britannica.com, Retrieved 19, June 2020

Permissions

All chapters in this book are published with permission under the Creative Commons Attribution Share Alike License or equivalent. Every chapter published in this book has been scrutinized by our experts. Their significance has been extensively debated. The topics covered herein carry significant information for a comprehensive understanding. They may even be implemented as practical applications or may be referred to as a beginning point for further studies.

We would like to thank the editorial team for lending their expertise to make the book truly unique. They have played a crucial role in the development of this book. Without their invaluable contributions this book wouldn't have been possible. They have made vital efforts to compile up to date information on the varied aspects of this subject to make this book a valuable addition to the collection of many professionals and students.

This book was conceptualized with the vision of imparting up-to-date and integrated information in this field. To ensure the same, a matchless editorial board was set up. Every individual on the board went through rigorous rounds of assessment to prove their worth. After which they invested a large part of their time researching and compiling the most relevant data for our readers.

The editorial board has been involved in producing this book since its inception. They have spent rigorous hours researching and exploring the diverse topics which have resulted in the successful publishing of this book. They have passed on their knowledge of decades through this book. To expedite this challenging task, the publisher supported the team at every step. A small team of assistant editors was also appointed to further simplify the editing procedure and attain best results for the readers.

Apart from the editorial board, the designing team has also invested a significant amount of their time in understanding the subject and creating the most relevant covers. They scrutinized every image to scout for the most suitable representation of the subject and create an appropriate cover for the book.

The publishing team has been an ardent support to the editorial, designing and production team. Their endless efforts to recruit the best for this project, has resulted in the accomplishment of this book. They are a veteran in the field of academics and their pool of knowledge is as vast as their experience in printing. Their expertise and guidance has proved useful at every step. Their uncompromising quality standards have made this book an exceptional effort. Their encouragement from time to time has been an inspiration for everyone.

The publisher and the editorial board hope that this book will prove to be a valuable piece of knowledge for students, practitioners and scholars across the globe.

Index

A
Alluvial Flat, 31-32
Applied Geomorphology, 4-5

B
Benthic Fauna, 15, 17
Biogeomorphology, 13-15, 95
Biological Abrasion, 15
Biostabilization, 14-17
Biota, 14-15, 18, 29, 96
Bioturbation, 14, 17

C
Carboniferous, 7, 9-10
Cartography, 3
Catchment Scale, 25
Cenozoic Era, 11
Chemical Weathering, 55, 164, 166-167, 169
Climatic Regions, 111
Composition, 14, 16, 18, 25, 35, 39-41, 45, 50-53, 55, 57-58, 60, 62, 110, 165, 206, 222
Computer Aided Cartography, 3
Coral Reef, 15, 226
Cretaceous Period, 10-11
Cyanobacteria, 15-16

D
Data Collection, 26, 96, 100, 107-108
Desertification, 124, 157, 181
Digital Elevation Model, 33
Digital Elevation Models, 3, 34
Digital Soil Mapping, 30

E
Electronic Distance Measurement, 3
Engineering Geomorphology, 5
Eocene Epoch, 11
Experimental Geomorphology, 5, 89

F
Fault Line, 213
Feedback, 14, 71, 96, 130, 211

G
Geographic Information System, 26-28
Geographic Information Systems, 26
Geological Structure, 2
Geological Time, 5-6, 8
Geomorphic Processes, 1-2, 24, 31-32, 39, 55, 58, 72, 81, 93, 110, 154
Geomorphologist, 2, 5, 19, 70, 94
Geomorphology, 1-5, 13, 15, 18-20, 22-24, 27-29, 33-35, 39, 70, 72-74, 77, 79, 81-86, 89, 93-95, 108, 125, 227
Geomorphometry, 3
Global Positioning System, 3

H
Hill Slope Scale, 25
Hydrogeomorphological Mapping, 27-28
Hydrogeomorphology, 24-29
Hydrological Cycle, 26, 29
Hydrosphere, 138

I
Ice Cap, 140
Ice Sheets, 135, 140, 173-174
Information Systems, 26

J
Jurassic Period, 10-11

K
Kinetic Energy, 41-42, 48, 128, 155, 186

L
Land Degradation, 124
Lava Plain, 231
Light Detection And Ranging, 2
Limiting Factor, 204
Loops, 208, 210-211

M
Magnesium, 55
Mesozoic Era, 10
Metadata, 31
Morphology, 2, 14, 18, 24-25, 39, 52, 81-82, 85, 87, 91, 95, 117, 125, 188, 197, 200, 202-203, 218-219
Morphometry, 19, 25, 35, 41, 77-79

N
Natural Landscape, 63

Non-spatial Data, 25
Northern Hemisphere, 11, 42

O
Ocean Currents, 196
Ordovician Period, 9
Orographic Effect, 2

P
Palaeozoic Era, 7, 11
Photogrammetry, 96
Physical Geography, 5, 175
Pleistocene Epoch, 12, 53
Precipitation, 25-26, 54-55, 128-129, 134, 136, 156, 173, 205

Q
Quantitative Revolution, 2

R
Radioactive Isotopes, 6
Radiometric Dating, 6
Remote Sensing, 1, 21, 24, 26-28
Resolution, 3, 31-33, 104

S
Satellite Imageries, 3, 26-27, 97
Sea Level, 41, 53, 115, 128-129, 137, 146, 186, 188, 196, 199-202, 205, 214, 219, 225-226
Sedimentary Rock, 62
Shield Volcano, 221
Simile, 92
Soft Substrates, 15
Soil Erosion, 20, 42, 44, 64, 93, 103, 105, 124, 156-157, 162, 188
Soil Geomorphology, 29, 33, 39
Spatial Data, 25, 28
Spatial Information, 27
Spatial Relationships, 31
Spatial Scales, 25, 30, 87, 99, 102
Stratovolcano, 222

T
Tectonic Uplift, 100-101, 186, 195, 214
Temporal Scales, 14, 87, 94, 99, 102, 104
Thermal Conductivity, 75, 165
Thermodynamics, 73
Time Scale, 5
Topography, 35, 39, 59, 78, 94, 125, 142, 153-154, 165, 193, 199-200, 204, 215-216
Triassic Period, 10

U
Urban Geomorphology, 18-20

V
Variables, 2, 71, 87, 94-98, 211

W
Water Budget, 26
Wave Action, 15, 55-56, 201, 219
Wave Energy, 14-16, 199, 201-202

www.ingramcontent.com/pod-product-compliance
Lightning Source LLC
LaVergne TN
LVHW080728210926
848485LV00003B/539